Healing Photons

The Science and Art of

Blood Irradiation Therapy

by

Kenneth J. Dillon

SCIENTIA PRESS

Washington, D.C.

Printed in the USA

January 1998

Publisher's Cataloging in Publication

Dillon, Kenneth J.
Healing Photons: the science and art of blood irradiation therapy / Kenneth J. Dillon.
180 p.; ill.; 22 cm.
Includes bibliographical references and index.
ISBN: 0-9642976-5-5

1. Phototherapy. 2. Lasers in medicine. 3. Ultraviolet radiation. 4. Blood--Transfusion, Autologous. I. Title.

RM842.D69 1998 615.831 97-062342

This book is for Dougie.

Nothing in this book should be taken as a claim that Blood Irradiation therapy is indicated as a treatment for any specific condition. No medical treatment should be administered solely on the basis of the information herein. Intending users should study the technical specifications, treatment modalities, and history of treatment of a specific indication before using any Blood Irradiation device. Special care should be taken to avoid exceeding the dosage limits of the Knott Technique.

Scientia Press is an imprint of Spectrum Bioscience, Inc.

Table of Contents

Introduction

Entia non sunt multiplicanda.

William of Ockham

"Occam's razor"--the admonition that entities are not to be multiplied--reminds us to strive for simplicity in science, and so it is appropriate here to point out and justify a major stylistic simplification characteristic of this book.

To delve into the obscure history and the arcane science of Blood Irradiation (BI) therapy[1] is to subject oneself to the temptation to evoke rhapsodic images of blissfully natural healing by means of gleaming rays of purest light. That shining light on human blood cells could transform them into a medicine that conveys excellent therapeutic benefit is surely a wonderful possibility. Such a medicine would endow humankind with a gift akin to the Philosophers' Stone.

Indeed, this possibility can be shown to exist in all of its inherent attractiveness. BI therapy is a wonder of nature and human art. This

---

[1] Blood Irradiation therapy has been termed photopheresis, Hematogenic Oxidation Therapy, Quantum Therapy, autotransfusion of UV-irradiated blood, and photoluminescence. In order to give BI the credibility it deserves among practitioners and patients who have never heard of it before, it would appear advisable to standardize on the terms "Ultraviolet Blood Irradiation" (UBI) or its foreign language equivalents and "LBI" (low-intensity laser blood irradiation). UBI is a simple, scientifically accurate acronym that has been used by practitioners over many decades to refer to the Knott Technique (see p. 8), while LBI is preferable to "low-intensity laser irradiation of the blood" or "weak" or "cold" laser irradiation of the blood.

Indeed, this possibility can be shown to exist in all of its inherent attractiveness. BI therapy is a wonder of nature and human art. This book explores the manifold dimensions of its allure. Yet to the best of my ability I have sought to steer clear of romanticism and hyperbole. Instead, this book offers fact upon fact, analysis upon analysis.

My justification? An unusual concept that flouts professional taboos is inevitably open to misinterpretation and criticism, so presenting it in the most sober fashion is essential. Thus this is a book of historical facts, clinical trials, discussions of applications, and scientific hypotheses and definitions. While that may rob the story of a certain superficial charm, the intrinsic fascination of the substance of Blood Irradiation therapy speaks for itself and requires no embroidering.

Before we proceed, two terms of primordial importance need to be marched front and center. They are:

1) as with every other therapy, the success of Blood Irradiation depends on dose; and

2) BI is not an *in vitro* experiment but an *in vivo* therapy that derives its potency from activated human blood cells.

Failure to keep these two elementary points in mind has led to many serious errors in the assessment of BI therapy.

Chapter 3 is drawn from the Internet BI Website: www.ubi.org. Certain sections of Chapter 3 are repeated in expanded form in the succeeding chapters on the various indications for BI therapy. In addition, instances of side effects are discussed at several points in the main text, then they are subjected to another examination in Appendix A. The perceptive Reader will recognize that this is a means of establishing credibility and in fact accords far more attention to BI's side effects than is merited by their infrequency and minor nature.

My thanks to Helen Bergan, Kusuma Cunningham, Kirill Denisin, Diane Foulds, Toni Herbert, Steve Marschalko, Michael Ryan, Gary Stein, Vladimir L. Voeikov, and my brothers Bob and Ray for their advice and support along the road to the Solar 1 device and the writing of this book.

And now to our story.

The Author

Imagination is more important than knowledge.

Albert Einstein

Chapter 1: Pages of History

Stretching back into the distant past, many human beings have looked to the sun as a source of healing. Physicians of antiquity, most notably Hippocrates, prescribed solar irradiation for various physical disorders. Sunbathing has been popular in many cultures since ancient times, and of course homo sapiens is not the only species that practices basking in the sun. Various cultures and individuals have also explored the potential healing properties of color, and elaborate systems of color therapy have evolved.

It was not until the 19th century, however, that advances in science allowed researchers to analyze solar radiation, define the characteristics of the solar spectrum, and begin to apply this new-found knowledge to medical ends. The first clinic for light therapy was established in Switzerland.[2] The Icelander Niels Ryberg Finsen's work on the medical uses of ultraviolet (UV) irradiation won him a Nobel prize. Another milestone was the discovery of the bactericidal action of UV radiation on microorganisms in fluids. The concurrent development of various UV light sources made laboratory and clinical use of UV possible.

Thus by 1920 extracorporeal light[3] therapy was in active use. As Dorno put it in 1911: "UV light is as healing as the surgeon's scalpel in the right dosage and as destructive of life in unconsidered use."[4] Researchers were busy studying ways of employing UV to purify water and other fluids. The popular use of color therapy was spreading, trailed by its scientific study. Scientists were not yet sure what the

---

[2] Anke Wiesner, "Zur Geschichte der UV-Therapie," Folia Haematologica, Leipzig 101 (1974), No. 5, p. 849

[3] Strictly speaking, of course, UV is not "light" in that it is invisible.

[4] Wiesner, ibid., p. 849

mechanisms of action of light therapy might be; but a growing body of evidence suggested that it was indeed effective, which was enough to ensure widespread and enthusiastic use.

Dr. Knott and Ultraviolet Blood Irradiation

It was a logical next step to apply the known bactericidal property of UV rays to irradiating the blood with the goal of destroying infectious organisms. This was first done by a researcher named K. Naswitis who shone UV on blood flowing through a shunt from a vein. As with other therapeutic approaches, of course, the real problem was not to try BI out but rather to study it in a systematic, scientific manner that would lead to the discovery of a demonstrably effective and safe way of applying this therapeutic approach. And with BI, that was a non-trivial matter, indeed.

In 1923 Seattle scientist (not M.D.) Dr. Emmet Knott began to experiment with ways to irradiate the blood to destroy infectious organisms.[5] It must be remembered that in an era without antibiotics, infectious diseases of all kinds were more widespread, far more devastating, and much less controllable, even in the most technologically developed countries. It was the fate of doctors to sit helplessly at the bedside of patients who acquired infections, often as a result of surgical procedures, whose fatal course no intervention could stop. So Knott was focussing on a very central problem of medicine.

Together with a medical student named L. A. Edblom, Knott constructed a crude apparatus that removed blood from a vein through a tube, citrated it to avoid coagulation, exposed it in a small irradiation chamber to calibrated UV irradiation, and then pumped it through a tube back into the body in a continuous process.

In experiments with dogs, Knott first attempted to irradiate the entire volume of blood after infecting the dogs so as to induce severe septicemia. He found that the irradiation cleared their blood of any trace of infection, but that they all died in 5-7 days of profound depression

---

[5] E.K. Knott, "Development of Ultraviolet Blood Irradiation," American Journal of Surgery 76 (1948), No. 2, pp. 165-71

and a progressive respiratory slow-up and failure. While it is hard to know exactly what had happened, in retrospect the symptoms are those of a massive overdose of chemotherapy.[6]

Undaunted, Knott persisted in trying different levels of irradiation on more dogs. On one occasion the apparatus failed partway through the irradiation of a dog, yet the dog survived, was free of infection, and showed no side effects. This led Knott to experiment with irradiating different amounts of the blood of the dogs until he determined that the optimal approach was to irradiate for about 10 seconds 1.5 to 3 milliliters of blood for every pound of body weight (perhaps 5 percent of the total blood) and to repeat the procedure after some days of waiting to observe the results of the first procedure. This approach came to be known as the Knott Technique and became the most widely used protocol. Ultimately, it would be shown by many researchers that the application of Ultraviolet Blood Irradiation (UBI) in repeated small doses is more therapeutically effective than the use of an equivalent single large dose.

The first treatment of a human occurred in 1928. The patient was a woman moribund following a septic abortion complicated by hemolytic streptococcus septicemia. Treatment with UBI rapidly returned her to normal health. Indicative of the caution and ethical probity with which Knott and his medical collaborators worked, they waited *five years* to make sure that the woman did not suffer from any subtle long-range effects before further treatment of a human subject in 1933 when the device again cured a patient with advanced hemolytic streptococcus septicemia. The UBI device then began to be used with some frequency on patients with severe septicemia and subsequently on patients with viral pneumonia.

Knott was thus the first to develop BI into a working therapy with clearly demonstrated dosage, method of application, therapeutic effect, and apparent safety. He never received any acknowledgement outside of a very narrow circle of initiates for what amounted to a major

---

[6] Decades years later, Russian researchers repeated Knott's experiment on 12 apparently healthy dogs. They reported that not only was there no harmful effect; the dogs became unusually energetic and lively but not aggressive after therapy. Three weeks after the experiment, a cataract disappeared from the eye of one dog. The dogs' red blood cells suffered no damage from the treatment, and bone marrow biopsies showed heightened leukopoiesis. V. I. Karandashov et al., "UV Irradiation of the Entire Circulating Blood [Russian]," Vestnik Rossiiskoi Akademii Meditsinskikh Nauk (1993), No. 9, pp. 38-41. Presumably these Russian researchers used a lower dose than Knott had used.

contribution to medical science. Knott patented the UBI device in 1928 along with Edblom.[7]

The Pioneering Years

In the late 1920s and early 1930s, various European physicians began to experiment with UBI. The Swiss Federico Wehrli began to use UBI on oxygenated blood in a procedure termed Hematogenic Oxidation Therapy (HOT) that has been used by hundreds of physicians in Central Europe and elsewhere ever since. The Czech Karel Havlíček independently hit upon the idea of using UBI in the 1930s, and others picked it up from him.

By the 1940s a handful of physicians in the US were regularly using the "Knott Hemo-Irradiator" according to the technique established by Knott. At first they concentrated on treating bacterial infections, but soon they expanded their use of UBI to deal with viral and other infectious diseases, inflammatory and autoimmune disorders, and ultimately a very wide range of disorders. They treated septicemias, pneumonia, poliomyelitis, botulism, non-healing wounds, encephalitis, peritonitis, asthma, pelvic inflammatory disease ("cellulitis"), biliary disease, hepatitis, and many other disorders.

There was some use by specialists, but the most dedicated proponents were general surgeons. At a time when a few sulfa drugs were just beginning to be used and otherwise there were no antibiotics available, these surgeons had a powerful need for something reliable to control postoperative infections that would kill their patients after they had successfully operated on them. In addition, as general surgeons, they had to deal with dozens of life-threatening emergencies and could not be concerned with the niceties of protocols and the qualms of armchair critics. They needed something that worked, and they found that UBI worked when nothing else did. And UBI appeared to be free of side effects other than flushing of the face in many cases and, rarely, chills from the citrate used as an anticoagulant. Also, the poor state of a

[7] The patent is in Appendix B.

patient's veins could cause trouble with extraction and return of the blood, as with any transfusion procedure.

The surgeons came to use UBI regularly both preoperationally and postoperationally. They and dozens of other physicians regularly used UBI. They reported their results in *The American Journal of Surgery* and other professional journals. Proponents of UBI founded the American Blood Irradiation Society.

A Few Controlled Studies

While most of the pioneering reports on UBI were lists of case histories, a few controlled studies were run. For instance, in 1944 G. J. P. Barger, M.D. of Georgetown University organized a comparative study of 25 cases of acute, severe bulbospinal polio in children, of which 11 received UBI treatment and 14 had standard medicinal treatment. Of the 11 receiving UBI, one died and the other 10 recovered completely. Of the 14 controls, 5 died and 9 recovered. In other words, UBI was much more effective.[8] Unfortunately, the trial was not reported in a regular medical journal.

In a controlled study in 1952 of 50 patients with acute or subacute iritis with or without iridocyclitis, UBI was used on 30 while the remainder received corticosteroids. All the UBI-treated patients recovered without complications or need of further treatment. In the control group 20 percent became blind in the affected eyes, and there were several complications, while even those who recovered did so more slowly than the UBI-treated group.[9]

Most of the evidence, however, was in the form of case studies. Thus, there was no way to verify whether the improvement noted was to be ascribed to UBI or to some other cause such as the delayed action of a drug administered before UBI, a spontaneous resurgence of the patient's immune system, or a psychological effect. In other words, the

---

[8] G.P. Miley. *Ultraviolet Blood Irradiation: A History and Guide to Clinical Application (1933-1997).* Silver Spring, Maryland: Foundation for Blood Irradiation, 1997, p. 33 (published posthumously)

[9] Ibid., p. 209

great majority of reports by the UBI doctors were suggestive and even persuasive, but they were not definitive or conclusive. Still, it is important to note that several controlled trials in the United States were on record by the early 1950s.

Reasons for Limited Acceptance

It may fairly be asked: if UBI was so safe and effective, why didn't more physicians begin to use it?

The reasons are interesting, complex, and important. One was that this era saw a proliferation of electro-therapeutic and other devices, many of them promoted by enthusiasts and charlatans, and hence often ineffective or downright dangerous. There was little government regulation of medical devices in the US and Europe, and many physicians felt it necessary to defend their patients against the ministrations of quacks.

Another reason was that the great discoveries in biochemistry had led medicine down the path of pharmacological treatments rather than biophysical ones. Few physicians had training in biophysics, and indeed biophysics itself was in its infancy.

A third reason might have been that the economic interest of the physicians and pharmaceutical companies strengthened existing prejudices against a therapy such as UBI that was relatively inexpensive and had a very wide range of application. There is no evidence pro or con on this issue in the record, and so it deserves only to be mentioned in passing as a possible explanation.

A fourth reason is that the proponents of UBI were ineffective in marketing their findings. They compiled long series of case histories of the treatment of various diseases with UBI. But frequently they did not include controls, and their criteria for success were often vague. Equally important, they deliberately refused to publicize UBI via the press or to make any special claims for it.

A fifth reason: the mechanism of action of UBI was very difficult to identify, which left them *championing, without the financial backing of a major pharmaceutical firm, a recently invented therapy consisting of invisible radiation that worked by some mysterious means, with purportedly wonderful results including virtually no side effects.*

Is it any wonder that they encountered skepticism and hostility?

Many of the early American proponents of UBI came under fierce attack by critical colleagues. Their reputations and careers suffered, deterring others from joining them. Still, they made enough headway by the 1940s that UBI had gained a certain minimal acceptance. A Blood Irradiation Clinic, headed by George P. Miley, M.D., was established at the Hahnemann Medical College Hospital of Philadelphia. Miley came to treat over 16,000 patients and wrote extensively on his findings. Among various leading pioneers was Robert C. Olney, M.D., who treated thousands of patients in Nebraska and wrote fundamental reports on the application of UBI to Pelvic Inflammatory Disease, biliary disease, and--along with 15 colleagues--hepatitis.

Antibiotics Take Over

Ultimately, the dramatic advent of the antibiotic revolution led many practitioners to set aside their UBI devices and made it difficult for the rest to publish in respectable medical journals. Antibiotics, accompanied by the new vaccines and corticosteroids, revolutionized the practice of medicine. They were wonder drugs, acclaimed by physicians and patients alike. Of course, in retrospect we know that microorganisms soon began to develop resistance to antibiotics, that these drugs could undermine the vitality of the immune system, and that the prices of some of them would eventually become very high. But in the 1950s antibiotics were understandably a sensational success.

The UBI practitioners were doctors, not master marketers. They also tended to be general practitioners and general surgeons, and so they did not concentrate on two or three indications in which UBI was clearly superior to antibiotics and vaccines.

In the US, the Olney *et al.* article on the use of UBI against hepatitis in 1955 can be seen as the last major concerted effort of UBI proponents to stem the tide.[10] In the conclusion to a study of 43 cases--three of them with severe chronic hepatitis successfully treated, with follow-ups of 4.5 to 6 years--the authors noted in a typically restrained manner

> In the face of the growing problem presented by this disease and the absence of any effective treatment for it, the results presented herein are considered significant and noteworthy.[11]

And of course they were right: antibiotics were worthless against hepatitis, while UBI showed great promise. One senses the frustration underlying their restrained language. It is as if they were asking, "Is anybody listening?" Evidently, nobody was. After that, the UBI practitioners tended to treat their own patients and steer clear of hostile state medical boards. In Europe UBI hung on, mainly in the form of Dr. Wehrli's HOT approach.

It is of fundamental importance to underline the fact that UBI was *not* discredited. No critic was ever able to find anything systematically wrong with UBI. It was simply abandoned by all but a few practitioners.

Knott had continued to work on upgrading the UBI device into the 1940s, with special attention to providing means such as curved grooving in the irradiation chamber to ensure that all of the blood cells received approximately the same amount of irradiation. Various companies put out imitation devices; but the "Knott Hemo-Irradiator", almost a cubic meter in size, remained the standard. Eventually it received FDA clearance as a "pre-amendment" or "510(k)" device that was sold and distributed in interstate commerce before 1976. Some of the devices ended up in museums, and others remained in use for decades.

---

[10] R.C. Olney *et al.* "Treatment of Viral Hepatitis with the Knott Technic of Blood Irradiation," American Journal of Surgery 90 (1955), pp. 402-9

[11] Ibid., p. 409

By the 1970s UBI came to be used by a growing band of East German physicians who eventually numbered as many as 400. The East Germans were particularly impressed by the excellent performance of UBI in the treatment of disorders of the microcirculation, e.g., of the lower extremities. They began to publish extensively on UBI. The best general handbook that emerged included a bibliography with 304 entries, most of them in German.[12] The Germans conducted many careful laboratory and clinical studies of UBI, but many of the clinical studies lacked strict controls and randomization. As with the earlier American studies, they tended to be series of case reports.

In Western Germany and elsewhere in Europe, a few physicians continued to use UBI.

The Russians[13], too, began to show interest in Blood Irradiation therapy in the 1970s.[14]

A number of conditions predisposed Russians to do so.

First, the lack of sunlight in the arctic regions of Russia caused scientists, physicians, and government to devote great efforts to the study of ultraviolet irradiation of the skin, leading to the development of impressive expertise in the medical uses of UV. Second, the high prestige of physics in the Soviet Union led naturally to an interest in biophysical approaches to medicine. Not only did Russian experts have excellent skills; Russian physicians had a considerably larger component of physics and mathematics in their training than their Western colleagues (this was also true in East Germany). Third, the high cost of pharmaceuticals prompted a search for non-drug therapies. Fourth, the rise of microbial resistance to antibiotics and the palpable shortcomings of chemotherapies in the treatment of circulatory and autoimmune diseases led to a search for non-drug therapies.

---

[12] Gerhard Frick. *Fibel der Ultraviolettbestrahlung des Blutes*. Greifswald: Ernst-Moritz-Arndt-Universitaet, 1989

[13] Strictly speaking, the word "Soviet" is more correct because researchers and physicians from many Union republics took part. In particular, the Ukrainians became actively involved. When we use "Russian", therefore, it should be understood that this is shorthand in many cases for "Soviet" or for a long list of the nationalities of the former Soviet Union. With apologies.

[14] Actually, Russian physicians had investigated it as early as the 1930s but had not pursued the subject.

It is legitimate to ask whether, in their eagerness to save money on healthcare, Russian medical bureaucrats were willing to overlook weaknesses in UBI therapy.

The answer is nuanced. Had UBI been somewhat less effective and safe than it in fact was, then eagerness to save money might indeed have led health officials to insist on its use even though it did not match the benefits to be obtained from expensive Western pharmaceuticals. The lack of money would have obliged medical doctors to go along since certainly treating patients with UBI was preferable to no treatment at all or to treatment with cheap, inadequate Soviet medications.

In fact, however, UBI was highly effective and safe; and so the situation was not so grim. Moreover, using the Isolda UBI device developed in Leningrad, Russian scientists and physicians subjected UBI to elaborate, painstaking laboratory testing early on, accompanied by scores of clinical trials. The Russians investigated UBI more thoroughly than either the Americans or the Germans had done, and with good reason. UBI was the perfect solution to their problems, it worked extremely well, and it opened up new vistas of medical care. Thus it was no wonder that they hunted assiduously for any shred of evidence regarding UBI's mechanisms of action or its effects in various indications, and they applied their superior knowledge of biophysics and specifically of UV irradiation to the task.

In addition, in the 1990s Russian physicians began to use low-intensity lasers to irradiate the blood through a fiber inserted into a vein with an IV needle (Laser Blood Irradiation: LBI). They found that LBI had very similar effects to UBI. Soon cheap, handheld helium-neon lasers operating at 632.8 nanometer wavelength and other wavelengths came to rival the UBI device and to a considerable extent to supplant it because of their relative ease of use. Many variants were tried, including the use of noncoherent light sources. UBI and LBI devices became standard features in Russian clinics and doctors' offices.

While there might have been an occasional physician in the pre-1990 era elsewhere in the world who used a UBI device, by far the main activity and all of the reporting occurred in Western countries, including Russia. In recent years, however, a few Chinese physicians have started to use BI and have made several reports of their results. The lack of usage of BI in the developing world is an anomaly because BI is ideally suited to the needs of these countries.

Thus the history of BI reveals a most curious and ironic pattern of a very remarkable and in many respects superior approach that, nonetheless, has fallen into disuse in its country of origin through no fault of its own. Not only that: many Western specialists in the treatment of such disorders as chronic hepatitis, necrotizing pancreatitis, bronchial asthma, and pelvic inflammatory disease are unaware that UBI was used with superior effectiveness and safety 50 years ago to treat these conditions. Hundreds of millions of dollars, if not more, are expended every year in searching for treatments for such disorders, and patients suffer from the lack of effective remedies. Yet long ago studies in respectable journals spelled out in considerable detail (if not with controls) the mastery that UBI could demonstrate over such disorders.[15]

It was a strange case of the loss of critical knowledge and lore--not of some obscure folk remedy from an impoverished backwater, but knowledge and lore developed right in the middle of the most technologically advanced country in the world. And now that BI has become identified with Russia, one more question mark is added to it in the eyes of Western practitioners.

---

[15] An abiding problem in this regard arises from the understandable caution of medical doctors in considering the adoption of a new therapy. They will bear the responsibility for any harm to their patients and so rightly insist on solid evidence prior to acceptance. However, they and many medical researchers fail to distinguish between evidence that would constitute "proof" and evidence that is highly suggestive of a therapy's efficacy and safety. The result is that they often consign an approach like BI to a category in which can be found all manner of dubious, unscientific, or entirely new therapies that lack the laboratory testing, clinical record, and logical underpinning that BI possesses in abundance. They also do not take into account the problem of financing studies of a therapy that is relatively inexpensive and so may not attract the necessary funding.

Chapter 2: How BI Works

Photobiology--the science of the interaction of light and living cells or creatures--is a dynamic, important area of biophysics. Certain parts of it such as the organization of the spectrum are quite straightforward and well-known; other parts, including the potentially therapeutic responses of cells to incoming light, are complex and little-understood. Among the complications: every kind of light and indeed every photon consists of electromagnetic radiation, and the persistence of this radiation over time constitutes an electromagnetic field. Thus when all the calculations of the effect of light at certain wavelengths on the chemistry of cellular structures are done, one must still ask whether there might be subtle magnetic and field aspects of the phenomenon that are not accounted for.

It must be assumed that over billions of years living cells have developed exquisitely attuned responses to solar irradiation. These might extend to aspects such as the degree of polarization of sunlight, the length of coherence of its wavelengths, changes in the sun's activity over short and very long terms, and radiation pressure.

Such considerations can lead to further questions. For instance, to what extent does an artificial source accurately represent the solar irradiation to which a living creature is accustomed, and what difference does it make? How significant are changes in the rhythm, intensity, and duration of irradiation? In what ways might the responses to irradiation of cells at least partially shielded from solar irradiation over eons of time (e.g., blood cells) differ from those of skin and eye cells?

We can assume that the main action of light in BI therapy is in the optical spectrum, approximately between 200 and 1,000 nanometers wavelength. This includes ultraviolet (UV-C: 190-290 nm; UV-B: 290-320 nm; and UV-A: 320-400, of which only wavelengths above 287 nm pass through the atmosphere without absorption); visible light

(400-700 nm); and infrared (from 700 nm up). X-rays and other electromagnetic radiation at farther reaches of the spectrum can also have therapeutic effects; but they are considered difficult to use in BI, potentially dangerous, and unnecessary in view of the effectiveness of the wavelengths in the optical spectrum.

Several laws of photochemistry govern these effects:

- the Grotthus-Draper Law (the first law of photochemistry) states that only light that is absorbed can act chemically. This law correctly assumes that in most cases the changes that occur in the chemistry of cellular components as a result of absorption of photons are what constitute the main photobiological effect. But it does not take into account such effects as coherence length, the radiation pressure of reflected light, and the possible influence of the electromagnetic fields of absorbed and nonabsorbed photons at ultra-low levels in the cells;

- the Bunsen-Roscoe Law (the second law of photochemistry) states that a photochemical change is proportional to the intensity and duration of illumination of the cells. Again, this is a useful first approximation; but in reality at times other factors come into play, including the rhythm of the illumination, the wavelength, the number of wavelengths, and the kind of cells (red blood cells react in a less predictable manner than reticular cells, for instance). Many researchers have found that lengthening the duration and lessening the intensity of irradiation can yield different results; and

- Stokes's Law states that the wavelengths of secondarily emitted radiation (i.e., the biophotons emitted by cells in response to the primary irradiation) are longer than the wavelengths of the primary irradiation. In BI this law correctly predicts that the secondary emissions from the irradiated cells are in the UV and visible areas of the spectrum, with the UV biophotons apparently at longer wavelengths than the initial UV irradiation.

In effect, these are all useful laws formulated in the early days of photobiology; but they are not cast in concrete, and the first two need to be revised in light of more recent theory, evidence, and instrumentation.

Biophotonic Radiation

Of particular relevance for BI is the body of theory and evidence that has developed in regard to Biophotonic Radiation (BR)--the ultraweak radiation emitted by all living cells--since its discovery by Alexander Gurvich in 1923.[16] In the theory of the cellular field that Gurvich developed and that has been elaborated by Fritz-Albert Popp and others, a deeper, more subtle stratum of activity emerges that underlies the common paradigm of molecular biology. Cells are connected in fields and collaborate according to information passed between them by means of biophotons. The coherence length of the incoming wave-signals determines the changes in these fields, which in turn stimulate or suppress biological activity. In some circumstances, the findings of the students of BR coincide with those of molecular biology; but where they diverge--e.g., in the study of regeneration[17] --the findings of the biophysicists appear to be more in tune with the observable evidence.

BI can thus be interpreted in terms of standard molecular biology, but it is at least equally appropriate to think of it as a form of *electromagnetism of the blood cells*--or as their energized mode. In turn, this concept explains some of the seemingly baffling aspects of BI.

Viewed from a systemic perspective, BI fits into a special branch of pharmacology--the pharmacology of Differential Energy Drugs (DEDs). DEDs are those pharmacologically active substances that derive their therapeutic effect from the degree to which they embody or appear to embody energy. The theory of DEDs links synthetic molecules such as nucleoside analogues; natural molecules such as ascorbic acid, glucose, and ATP; and various kinds of Blood Irradiation and related therapies in an ascending ladder of specificity according to the degree to which they represent the biological ideal of energy, which is solar irradiation. The principles governing the photobiological DEDs

---

[16] See Marco Bischof, *Biophotonen: Das Licht in unseren Zellen*. Frankfurt: Zweitausendeins, 1995; and Mae-Wan Ho. *The Rainbow and the Worm: The Physics of Organisms*. Singapore: World Scientific, 1993.

[17] See Robert O. Becker, *The Body Electric: Electromagnetism and the Foundation of Life*. New York: Morrow, 1985.

are discussed in the answers to the Frequently Asked Questions in Chapter 3.[18] In effect, they form a kind of biophysical pharmacology.

The Hunt for BI's Mechanism of Action

When Knott and others set out to harness the known bactericidal activity of ultraviolet irradiation to destroy microorganisms in the blood, they assumed that the most effective UV would be UV-C around 254 nm, since that was the most effective in destroying bacteria in other circumstances.[19] Knott found, however, that his device was only partly effective in destroying bacteria in the blood with its direct irradiation, yet the clinical effects were much better than this partial destruction would have led to. So he concluded that the procedure triggered a more complex effect once the irradiated cells returned to the body. He and others identified this with the action of the secondary radiation and the more general improvement of the immunological defenses of the body that emerged following the procedure.

Unfortunately, researchers and medical doctors who were not aware of Knott's finding and revised hypothesis continued to assume that the results claimed for UBI were obtained via the direct irradiation of microorganisms with UV-C. When critics of UBI tested the device on bacteria in the blood and examined samples of irradiated blood before returning it to the body, they "discovered" that the destruction of the bacteria was only partial, and as a result they tended to dismiss UBI as ineffective.

---

[18] See also Kenneth J. Dillon, *Apprentice to Paracelsus: My Search for the Secrets of Healing*. McLean, Virginia: McLean Research Associates, 1994.

[19] In fact, many combinations of UV wavelengths have been used in different devices. Miley, for instance, used a later version of the Knott Hemo-Irradiator with a main wavelength of 365 nm, i.e., UV-A, though it also had significant outputs in UV-B and UV-C. The Russian Izolda model now used has a main wavelength of 254 nm, i.e., UV-C, though its other wavelengths extend through the UV part of the spectrum into the visible part. While there is considerable debate over which wavelength has the greatest therapeutic effect, much evidence points toward the conclusion that they all--including visible light, though perhaps not infrared--convey approximately the same effect.

A very prominent feature of the history and current status of BI is the propensity of many observers to apply to a new, unusual phenomenon some formulaic test or *ad hoc* criterion that is based on a faulty assumption (in this case, confusing *in vitro* with *in vivo* mechanisms). When they find evidence, however tenuous, that supports their reservations, they dismiss the new phenomenon out of hand, without questioning whether their test or criterion was appropriate or why other seemingly intelligent, reputable, and more experienced colleagues have arrived at a different conclusion.

Photobiology 101

Why was it and is it important to know the mechanisms of action of BI? After all, many therapies, including natural ones, seem to work quite nicely even though we don't know exactly why.

One reason for wanting to know why BI works is scientific curiosity. If we come to understand something mysterious, it is enlightening and satisfying. There may also be practical implications for other aspects of medicine. But the overriding motive for investigating BI's mechanisms of action has been to gain assurance that indeed it does no hidden damage to the body--and thus to provide a cause for physicians to feel trust in using it. An understanding of the mechanisms of action of a given therapy is second only to rigorous clinical trials with clearly positive results in its power to lead physicians to accept a given therapy.

If a single truth has emerged from all of the many studies of BI's action, it is that there is no one mechanism of action of BI. Its action is complex. Indeed, while it is easy to detect the various components of the body's response to BI, it is much harder to define which of them is primary and which are secondary. However, careful analysis seems to show that certain ones are indeed more basic than others, or so it can be argued.

BI's pattern of action runs parallel to that caused by UV irradiation of the skin. According to Russian researchers, in both cases there are:

> 1) a broad spectrum of medicinal-healing effects (general strengthening, desensitizing, stimulating, anti-inflammatory) in various illnesses: inflammatory, infectious, cardio-circulatory, dermatological, autoimmune, and others; 2) the stimulation of factors of non-specific defense and immunity; 3) anti-hypoxic, vasodilatory effects, improvement of the rheological characteristics of the blood and the microcirculation; 4) activation of metabolic processes and improvement of their regulation; 5) stimulation of proliferative processes--hematopoiesis and regeneration; 6) rapid and to a great extent concurrent changes in cells and blood plasma; 7) a regulating (normalizing) kind of effect, including in the stimulation of only those processes, elements, and functions which are below the normal level; if the corresponding indicators are high, one observes a lowering or disappearance of their influence; 8) a direct dependence of the kind of effect of the irradiation on its dose, i.e., there is a stimulating effect of small doses and an inhibitory effect of large ones.[20]

Rays longer than 290 nm (and this means virtually all rays in the optical spectrum on the Earth's surface since UV under 287 nm is absorbed in the atmosphere) penetrate sufficiently deeply to affect the blood immediately beneath the skin. Many of the health benefits of solar and artificial UV irradiation are thus mediated by the blood. However, the therapeutic window for this action is quite narrow. To obtain benefits that would rival those available via BI would run the risk of severe damage to the skin; obtaining sufficiently precise and large doses is quite difficult, though in theory it is possible.

In BI, the incoming photons are--with the exception of a few in the UV-C range--of sufficiently long wavelength and low frequency that they do not cause ionization of the atoms they reach.[21] They lack the energy to knock electrons totally free from the atom. According to the laws of quantum physics and the arrangement of shells of electrons

---

[20] I.E. Ganelina and K.A. Samoilova, eds. *Mechanisms of the Influence of Blood Irradiated with Ultraviolet Rays on the Organisms of Humans and Animals* [Russian]. Leningrad: Nauka, 1986, pp. 155-6

[21] See J. Jagger, *Introduction to Research in Ultraviolet Photobiology*. Englewood Cliffs, New Jersey: Prentice-Hall, 1967, from which the following discussion is largely drawn. See also Thomas P. Coohill, "Action Spectra Again?" Photochemistry and Photobiology 54 (1991), No. 5, pp. 859-870.

around the nucleus, photons with less than ionizing energy can excite electrons only if they have just the right energy to raise them to a permitted energy level (excited state). This accounts for the discrete absorption spectra.

When the blood is irradiated in BI, an incoming photon (UV in UBI, red in the most common form of LBI, and varying wavelengths in secondary radiation emitted by irradiated cells) can excite an electron to move to a higher orbit in the electron shells of an atom. Upon dropping back to its normal place, the electron then itself emits a photon. Depending on how high the initial excitation is and the steps the electron takes as it drops back from one shell to another to its "ground state," the result can be fluorescence, phosphorescence, or what Gurvich termed "degradative" radiation. The new biophoton is then free to continue the process with other electrons, generally at a lower level of energy/longer wavelength (Stokes's Law).

Most molecules in solution (as in the body) do not immediately re-emit a photon upon absorbing one. Instead, there may be *direct utilization* involving breakage of a chemical bond, oxidation, or reduction--almost always in gaseous states. Another pattern is called *stabilization*, i.e., the energy is held for later use, for instance via a *cis-trans* change. The third, most common reaction is *degradation*: an excited molecule in collision with another converts some of its energy into translation, vibration, or rotation of the other molecule. The resultant emitted photons are then of lower energy. Finally, so-called *transfer* can occur: on an intramolecular, intermolecular, or "resonance" (i.e., radiationless transfer across fewer than 5 nm) basis, energy is transferred to other parts of the living tissue.

An important result of these complex reactions following absorption is that various kinds of vibration and rotation tend to blur the spectral structure of emissions of biophotons. Thus in BI the spectra of secondary emissions of biophotons are very broad.

When the many complex reactions that follow absorption, the irregularities in size and shape of cells, and the differential penetration of irradiation into a biological specimen are taken into consideration, it becomes very difficult to ascertain a simple "action spectrum." Instead, one needs to specify the organism, the effect, and the conditions. And even then the exercise is fraught with uncertainty. While photobiologists strive to attain precision, the complex structures and dynamics of living matter conspire to thwart them.

Among organic molecules, proteins are quite sensitive to UV irradiation, which can affect the bonds of certain aromatic amino acids. The disulfide bond in cystine maintains the tertiary structure of many proteins, making cystine an important target for UV. Peptide bonds are also key targets because they occur so frequently.

Nucleic acid residues (bases), especially pyrimidines, are even more vulnerable to UV than proteins. This means that thymine and cytosine in DNA and uracil and cytosine in RNA are vital targets in which UV or secondary biophotons concentrate.

When oxygen molecules are involved in the reactions resulting from BI's secondary radiation's biophotons, the result is an oxidative burst--a branching chain reaction of oxygen molecules that releases oxygen radicals. In particular, the hemoglobin in red blood cells, as the carrier of oxygen in the blood, plays a key role. The free oxygen radicals in turn set off a complex series of actions in a large number of blood constituents that leads to dramatic changes. A consideration of BI as an oxygenating therapy can lead into a broader discussion of BI's mechanisms of action.

Oxygenation

BI has a powerful oxygenating effect. Within minutes of the return of the irradiated blood to the body, venous oxygen levels rise. Miley reported an average rise of 58.6 percent after 10 minutes in 58 patients. This is independent of the numbers of erythrocytes or the level of hemoglobin, neither of which register changes that could account for this rise. The lower the oxygen level is to start with, the greater the rise tends to be. While venous oxygen then, seemingly unaccountably, dips after a half-hour, it eventually tends to become and remain higher for weeks after the procedure.[22]

A consequence of this is that cyanosis tends to clear up rapidly, especially in cases of pneumonia; and the more general perfusion of

[22] Miley (1997), p. 20; G.M. Miley, "The Ultra-violet Irradiation of Auto-transfused Human Blood: Studies in Oxygen Absorption Values," American Journal of Medical Sciences, June 1939, p. 873

tissues with oxygen cannot help but be of general benefit. For this reason, various physicians who practice oxygenation therapies view UBI as an important tool. (BI has also been used to oxygenate the blood of athletes, a dubious application.)

Of equal interest, changes in metabolism that begin very soon after the reinfusion of the irradiated blood lead to a significant improvement in the ability of the local tissue to extract the oxygen from the blood. In lower extremity circulatory disorders, patients at times will report that their feet feel warm even though the arteries remain occluded, and this effect may explain the otherwise unaccountable dip in venous oxygen reported.[23]

A related rapid effect of BI is vasodilation, which in turn eases the flow of blood and the uptake of oxygen by the tissues. In some cases this vasodilation causes a flushing of the face.

Both oxygenation and direct BI action on the nerves of the autonomic nervous system give rise to a general relaxation effect. This can be seen in the resolution of paralytic ileus, in the relaxation of the sphincter of Oddi that helps in the treatment of necrotizing pancreatitis, in the relaxation of bronchial muscle in bronchial asthma, and in many other areas.

A small amount of ozone is formed in the course of the oxygenation process. It is not clear whether this ozone has any deleterious effects.

Effects on Blood Components

Of the blood cells, activated platelets appear to be most sensitive to doses of BI. Relatively modest doses can suppress their activity; larger ones can destroy them. Fibrinolytic factors released under the impact of the secondary radiation tend to dissolve clots, hastened by a decline of fibrinogen, thrombin, and prothrombin. Nonetheless, the great majority of platelets--those that are not excessively activated--appear to benefit from BI. When they are deformed as a result of disease processes, BI reshapes their ultrastructure into a normal pattern. The

[23] Frick (1989), pp. 24-5

serotonin in platelets gives off a faint luminescence following BI, while the platelets themselves can give off a strong luminescence in plasma.[24] The platelets also develop greater electrophoretic mobility.

Of the white blood cells, T-lymphocytes are somewhat sensitive, B-cells less so, and granulocytes and macrophages least of all.

Within one hour of the procedure, all red blood cells in the circulating blood register changes in their membranes, including the expression of antigens. It is also possible that the red blood cells excrete biologically active substances. There may be concomitant heightened autoimmune hemolysis (destruction of blood cells).[25] It is possible that the direct beams of LBI are somewhat more destructive of red blood cell membranes than UBI.

BI dramatically improves the rheological (flow) characteristics of the blood, as much as by 50 percent or more, accompanied by a drop in blood viscosity. The tendency of red blood cells to aggregate is sharply reduced as is their deformation. Russian researchers attributed these effects to changes in the electrical and osmotic characteristics of the cell membranes; similar changes take place in platelets.[26] Surface antibodies on red blood cell membranes of patients increase by as much as 8-16 times in response to UBI.

Decreases in blood volume, diastolic and systolic blood pressure, and hematocrit also occur.

UBI increases the activation of circulating IgM (by 2-16 times), IgG (by 2-4 times), and complement (by 2-4 times), i.e., it rapidly activates these factors of nonspecific immune defense.[27]

BI causes structural changes in blood proteins such as hemoglobin, ceruloplasmin, and albumin in addition to IgM, IgG, and complement. Although their spectral absorption maxima are at 210 and 280 nm, this is of little assistance in determining which wavelengths of UBI might be most effective because its action on them is mainly indirect, mediated via the secondary radiation emitted from the irradiated blood cells upon their return to the body and in concentrations that are hard to measure.

BI reduces blood cholesterol and improves the balance between HDL and LDL cholesterol.

---

[24] I.N. Piksin et al., "UBI in Surgery [Russian]," Khirurgiia (1990), No. 11, pp. 100-4

[25] Ganelina and Samoilova (1986), pp. 165-6

[26] Ibid., pp. 207-11

[27] Ibid., p. 236

Small doses of BI activate phagocytes. BI causes pH in the phagosomes to drop to an acidic 5.0, which is lethal for certain microorganisms; and this pH level is optimal for hydrolase and to some extent acid phosphatase, also destructive of bacteria.[28] BI also leads basophile granulocytes and mast cells to degranulate, releasing heparin, histamine, and catecholamines. In larger doses, BI suppresses the immune function of white blood cells and destroys a portion of them.

The oxidative chain reaction set off by the reinfusion of the activated blood cells leads to a brief surge in lipid peroxidation, which in turn alters the membranes of blood cells. At the same time, however, BI also boosts the antioxidative capacity of the blood by activating peroxidase. Russian researchers reported that after a second or third session of UBI, this activation reached 40-50 percent above the initial level.[29] Thus the longer-run effect is to stabilize the membranes of red blood cells against the lipid peroxidation that can otherwise be so damaging in such syndromes as preeclampsia.

Meanwhile, the entry of the secondary radiation into the bone marrow stimulates hematopoiesis. The new blood cells migrate over the days following therapy into the peripheral blood and exercise a regulating, normalizing, and optimizing function there that supplements the initial effects of BI on the circulating peripheral blood. BI also activates hormones and stimulates their formation.

The combined action of all these effects and others is to change the correlation of forces, so to speak, in the blood. Many studies have shown that BI has a powerful regulatory function, i.e., it suppresses excessive activity and stimulates underperforming elements, so that a general normalization of the numbers and functioning of the blood components emerges. The more a given factor is out of line at the outset, the more decisively BI tends to act on it.

In addition, the very process of extracting the blood from the body, even without irradiation, is known to convey certain of the beneficial effects of BI. Presumably the blood cells absorb ambient photons, which act on them like BI. Autologous transfusion is a recognized therapeutic approach, including by many physicians who have never heard of BI.

---

[28] Piksin (1990), p. 102

[29] Ibid.

The Underlying Mechanism

Knott and his associates identified two possible modes that would explain how BI therapy obtains its effects:

1. the UV irradiation of the blood in the irradiation chamber destroys or alters bacteria and viruses in the extracted blood in such a way as to create a kind of vaccination effect when they return to the body. This provokes a reaction by the immune system which in turn destroys most or all of the other bacteria or virus in the body; and
2. the irradiation of a small fraction (some 5 percent) of the blood then spreads throughout the entire volume of the blood upon returning to the body, and this induced secondary radiation destroys virus, bacteria, and--in autoimmune diseases--activated white blood cells.

It is also possible that the fragments of bacteria, virus, and cells that are destroyed by BI act as a kind of vaccine in the plasma, enhancing the immune response.

Knott noted that BI treatment has a complex effect on the immune system. On the one hand, BI stimulates the activity of white blood cells; on the other, excess amounts destroy various white blood cells. The first effect is the basis of the immune response explanation of the beneficial results of BI treatment. The second suggests a reason why BI treatment can be so effective against autoimmune diseases. In these disorders, the metabolically active T-cells and other immune cells absorb much greater amounts of radiation than ordinary body cells, and this radiation inhibits them or destroys them, thus slowing down or stopping the disease.

Activated T-cells in particular are prone to absorb the induced secondary radiation following BI treatment as a source of energy just as they absorb at a very high rate glucose and other energy-bearing molecules. In effect, they are tricked by evolution. Having specialized for hundreds of millions of years within the controlled environment of the bodies of animals in the art of absorbing as much endogenous biochemical energy as possible in their activated state (via the "glucose shunt" a cell can absorb over 1,000 molecules of glucose per second), they are not equipped to switch to shutting out excessive energy that comes without warning from outside the body.

An Energy Gradient

The remarkable specificity that BI treatment demonstrates can best be explained by the body's own system of shuttling energy around to the places it is needed. This effect can be seen most readily in the fulminating conditions against which BI has shown itself to be so formidable. These conditions--e.g., fulminant hepatitis--suck into themselves an unusually high amount of energy in the form of glucose and other energy-bearing molecules. As the fulmination spirals upward, the body smoothly fuels it with energy, suggesting that there is a kind of Energy Gradient in the blood--a system whereby the body supplies energy to the various processes in it on demand and, if necessary, to a far higher degree than would occur by the mere undirected circulation of energy-bearing molecules via the blood.

This Energy Gradient explains the exceptional specificity of BI treatment in fulminating conditions: in effect the bloodborne secondary radiation is channeled as energy directly toward the fulmination, where it destroys the activated immune cells (or, in the case of necrotizing pancreatitis, the activated enzymes) that are driving it. In these circumstances, even amounts of radiation from BI treatment well over the normal dosage tend to do little or no peripheral damage, in contrast to treatment with various chemotherapies.

Another possible explanation of the effectiveness of BI treatment in the special case of liver diseases is that the blood-filtering action of the liver tends to concentrate the radiation to a far higher level than the modest levels in the circulating blood. This effect would suggest that BI might be equally effective in the treatment of Idiopathic Thrombocytopenic Purpura (ITP), an autoimmune disease of the spleen, another blood-filtering organ.

In addition, as a fluid the blood is capable of delivering the secondary radiation from BI to hard-to-get-at locations in the body which other kinds of radiation cannot reach without damaging tissue.

The result is high specificity This explains the action of BI in neurological disorders such as *petit mal* seizures. While T-cells are not implicated in these neurological and psychiatric disorders, BI simply substitutes for the glucose that fuels whatever metabolically active cells or processes are involved. In this sense, BI can be seen as a kind of glucose antagonist/substitute/overrider and thus as a suppressor of any excessive metabolic activity in the brain--or for that matter anywhere else in the body.

In contrast, in the lower concentrations with which BI affects cells, enzymes, and other factors that are underperforming in certain disease states, e.g., fibrinolytic elements in arteriosclerosis, BI can have a stimulating effect. Thus its overall action is to normalize the situation by suppressing excessively active factors and stimulating underperformers. A single dose of BI can therefore be both "immunostimulatory" and "immunosuppressive" depending on which sets of cells are under discussion. Likewise, an initial dose of BI can stimulate a cell, but then repeated doses can eventually inhibit it or destroy it. Thus the fact that rheumatoid arthritis or bronchial asthma may respond to initial doses of BI with a typical flare-up generally means that subsequent doses will be highly effective in suppressing the very cells that cause the flare-up.

Once BI inhibits or destroys cells with excessive metabolic activity, glucose that would otherwise flow to them becomes available to underactive cells, which enhances the normalizing effect.

The immunostimulatory, normalizing (i.e., optimizing) property of BI stands in marked contrast to the action of antibiotics and other chemotherapeutic approaches, which tend to disrupt the body's mechanisms and lead to imbalances, including among flora. There is no evidence that BI affects the body's flora except in the case of an infection, when it destroys the overly active, excessive bacteria.

In certain disease states it is possible that double- and triple-concentration effects occur, and these can powerfully boost BI's specificity. For instance, in a case of fulminating primary biliary cirrhosis, the initially mild level of secondary radiation in the blood could be concentrated in three ways: by the filtering action of the liver, by the blood's Energy Gradient, and by the differential absorption of the radiation by the activated T-cells. These effects would not merely be added to each other; they would be multiplied by each other, leading to an exceptional specificity that would explain why a very modest amount of extracorporeal BI can have the dramatic localized effect that it does.

It should also be noted that BI can be accompanied by psychologically-induced effects akin to a placebo effect and arising from patients' perception of it as especially powerful or as more natural and less toxic than chemotherapy. This effect has been documented in many studies. In a clinical trial of LBI in rheumatoid arthritis, for instance, a control group of 18 was given an "intrusive placebo" that consisted of the daily insertion of the IV laser waveguide, except that unbeknownst to them the laser beam was not turned on. Two patients had "significant improvement" and 12 had "improvement", which equalled the effects achieved by another group that received 4-6 sessions of LBI over the course of two weeks--though it was nowhere near as good as the response of two groups that were treated daily for six days.[30]

Overall Assessment

In sum, BI therapy operates in a somewhat complex manner but frequently with a surprisingly simple specificity and consequent virtual lack of side effects. In infectious diseases the immunostimulatory effect and the induced secondary radiation work in tandem. In autoimmune disorders, the concentrated secondary radiation appears to be the primary or even sole mode by which BI treatment obtains its effects, suggesting that even in infectious diseases it plays a more important role than the immunostimulatory effect.

The hypothesis that the induced secondary radiation is the most important mechanism of action of BI fits perfectly the pattern of damage that unusually high doses of BI can do as well as the known pattern of specificity of antimetabolite drugs (e.g., 2-CdA/Cladribine) that resemble BI in the sense that they mimic energy-bearing molecules and thus to some extent serve as glucose antagonists. BI can be viewed as the single most powerful of the antimetabolites.

The method of action of the energy-bearing secondary radiation from BI treatment appears to lend it higher specificity than many chemotherapies aimed at the same applications, since to attain their

[30] K.V. Zvereva et al., "The Choice of Method of Intravascular Laser Therapy in Rheumatoid Arthritis [Russian]," Terapevticheskii Arkhiv 66 (1994), No. 1, pp. 29-32

effects such chemotherapies must deviate from the ideal purity of energy-bearing molecules such as glucose and ATP.[31] This suggests in turn that the negligible observed side effects of properly administered BI treatment are not the tip of an iceberg of hidden damage but rather that BI indeed has exactly the exceptionally high specificity that the above energy-bearing model implies. In other words, BI therapy is not only safe; it is safer than competing chemotherapies throughout a wide band of therapeutic action.

The dramatic advances over the past 50 years in medical science's understanding of physiology and in its ability to monitor disease states mean that it will now be possible to achieve a better grasp of how BI therapy works. In turn, this will provide insight into various physiological processes as well as an opportunity to fine-tune the therapy and apply it to new indications, thus achieving significantly better results than the pioneers of BI treatment were able to obtain.

---

[31] Dillon (1994), pp. 37-45

Chapter 3: Frequently Asked Questions about BI

It is natural at this point that the Reader will have various questions about BI. So it makes sense to try to pose them and then answer them. The non-American Reader's indulgence is requested for the emphasis on the American situation.

What is the regulatory status of the UBI device?

The UBI device has US Food and Drug Administration (FDA) clearance as a medical device that was in interstate commerce prior to 1976. This is not the same as "FDA approval", which would require demonstrating that it is effective and safe for specific indications by means of controlled clinical trials. But the UBI device may be marketed for use by licensed practitioners in the United States without any claims being made regarding specific indications. It is unclear whether the Russian Isolda model meets FDA standards. The FDA has approved the principle of BI therapy (see p. 48).

Do insurers cover treatments with BI?

Some do.

Who is authorized to use a BI device?

In the United States, any licensed medical practitioner is authorized to use a UBI device. It is ideal, for instance, for use by a registered nurse with an independent practice who accepts patients referred by physicians for BI treatment.

Does that include Blood Irradiation via low-intensity laser devices (LBI)?

Although LBI has become very popular in Russia, the use of low-intensity laser devices for Blood Irradiation has not yet spread to the United States. It is therefore not clear whether any practitioner is permitted to administer LBI. In reality, however, it would appear difficult to regulate LBI because lasers have become such accepted and ubiquitous devices. Low-intensity lasers are rapidly gaining acceptance in countries like the UK for local treatment of pain and wound-healing. Veterinarians in the US use them to treat a range of ailments, particularly of horses. It is just a matter of time before their much broader application via BI therapy becomes known.

Which is better: UBI or LBI?

The therapeutic effects of these two modes of BI are similar. In a study of the treatment of 312 workers who had received significant doses of radiation during the cleanup of the Chernobyl nuclear accident, UBI was used on 54 and LBI on 126, with 132 receiving standard pharmacological treatment for a range of disorders: vegetative dystonia, dyscirculatory encephalitis, hypertonic disorder, gastrointestinal disorders, chronic hepatitis, and chronic bronchitis. A normalization of microcirculatory and immunological indicators occurred in 73 percent of the UBI cases and 84.8 percent of the LBI cases. But 39 of the LBI cases received an extra drug as well, and no follow-up tests were done to identify delayed effects.[32]

Clinical trials in Vladivostok of the comparative effectiveness of UBI and LBI in conjunction with fasting in the treatment of hundreds of patients with bronchial asthma yielded a nuanced but highly interesting result.[33] No differences were found in the effects on bronchiectasis.

---

[32] V.M. Frolov et al., "Compromised Immunity and Microcirculation and Their Correction in the Liquidators of the Aftermath of the Chernobyl' Reactor Accident, Presenting with Neurosomatic Pathologies [Russian]," Likarska Sprava (1995), No. 5-6, pp. 22-25

[33] G.I. Sukhanova, ed. *Laser Therapy in the Far East* [Russian]. Vladivostok: Dal'nauka, 1993, pp. 67-68

LBI had a more rapid effect overall and was superior in terms of bronchodilation and hyposensitization, while UBI had a more marked antibactericidal and anti-inflammatory effect. Two weeks after treatment, however, UBI obtained better results in terms of microcirculation as well. LBI brought a rise in the absolute number of lymphocytes and the number of helper T-cells as well as a decline in the number of B-cells and the phagocytic activity of neutrophils, whereas UBI lowered the absolute number of lymphocytes, the number of B-cells, and the phagocytic activity of the neutrophils.

The researchers concluded that LBI's greater ease of use and more rapid effects made it superior for less serious cases of bronchial asthma, while UBI was to be favored for more serious cases with infectious features. They noted that either approach combined with fasting conveyed a "beautiful therapeutic effect."

The trials did not test the comparative merits of UBI or LBI as a stand-alone therapy versus their use in conjunction with fasting. In addition, the specifics of bronchial asthma may not pertain in other indications. Moreover, the differences between UBI and LBI were not great and may have arisen from unintended differences in dosage, though the fact that the curves of the results of LBI and UBI crossed each other as UBI forged ahead in the second week after treatment suggests that the dosages were roughly equivalent. UBI's superiority also showed up in four successive trials on bronchial asthma in hundreds of patients--a powerful indication that in fact it has a more profoundly therapeutic effect.

The reported advantage of UBI in treating severe cases may not have been due to the known bactericidal and virucidal effect of UV irradiation at all because relatively few cells were directly irradiated with UV. Three other possible explanations are: 1) the greater number of wavelengths in UBI stimulated the irradiated cells in a more intense, "complete" way reminiscent of solar irradiation than did the single wavelength (632.8 nm) of the red laser; 2) the addition of ambient photons in the extracorporeal mode of UBI might have provided greater stimulation; and 3) UBI had a more profound impact on the marrow. The delayed effect of UBI could have arisen from the production of more cells in the marrow over the course of 10 days. The effects of BI in the marrow are little studied and could be an important source of insight.

In assessing the UBI vs. LBI question, various considerations are apropos. Clearly, the Vladivostok report deserves much more credence than the Ukrainian one. The Vladivostok researchers had specialized in UBI for many years; one of the first books on UBI was published in Vladivostok in 1987. Their methodology and execution were much more scientific. They also clearly stated their initial preference for using LBI to the exclusion of UBI because it was a less cumbersome, more precise approach. This gives higher credence to their conclusion that in fact UBI is somewhat more effective than LBI. Their ascription of this superiority to the germicidal effect of UBI is, however, almost surely incorrect. The fuller spectrum and marrow effects of the UBI device appear to be much more persuasive explanations.

In general, LBI has the advantages that the hand-held device is easier to use and the amount of irradiation can be precisely calculated. It does also not require an irradiation chamber, so there is no need for cleaning quartz irradiation chambers or paying for disposable ones. It also saves the trouble and expense of using heparin as well as problems with coagulation.

The UBI device's advantage is that it can be used by any individual with basic medical training, thus saving the expense of a professional's time and permitting its use in situations where a professional is not present. UBI has the advantage, too, of all extracorporeal blood treatment that the blood also absorbs ambient photons that can have a beneficial effect once it returns to the body.

How is LBI administered?

Many approaches have been tried. A particularly effective one is to use in 30-minute sessions a helium-neon laser with 1-5 milliwatt output at the tip of a polymer-coated quartz fiber (ranging from 200 to 600 micrometers in diameter) inserted into a vein with an IV needle, which is then withdrawn. For serious chronic disorders like rheumatoid arthritis and schizophrenia, the sessions are repeated daily 5-15 times, depending on the patient's condition. Raising the wattage seems to convey little or no benefit. As with UBI, repeated small doses have better cumulative effects than the same total dosage administered in a single session.

For certain local conditions, LBI appears to be more effective than local low-intensity laser therapy. A study of the effects of the two approaches in treating degenerative diseases of the lower extremities concluded that LBI was distinctly superior to local laser therapy.[34]

[34] G.E. Brill', ed. *The Experimental and Clinical Use of Low-Intensity Lasers and Irradiation in the Millimeter Range* [Russian]. Saratov, 1994, p. 148

According to another study, LBI via transdermal irradiation of a vein with an infrared laser of a longer wavelength (830 nm vs. the normal 632.8 nm--the longer wavelength permitted deeper penetration in order to irradiate the blood more effectively) provided decent results while avoiding the invasiveness of an IV needle and fiber. But the results were not nearly as good as with IV LBI.[35]

A study of 980 nm transdermal LBI administered in several spots around the heart to patients with ischemic heart disease, however, found that the results were somewhat superior to those of intravascular LBI in diminishing blood cholesterol and other blood lipids for up to six months.[36] It would be interesting to compare this approach with UBI.

A series of St. Petersburg studies of the use of transdermal LBI have reportedly yielded good results on chronic hepatitis and other disorders, though there was no direct comparison with other forms of BI therapy.[37]

One concern in regard to IV LBI is that the fiber waveguide, which commonly extends 2-5 cm downstream in the vein from the point of insertion, could become a source of transmission of infection to other patients. The careful cleaning procedures used in Russia and Ukraine are still not sufficient to overcome this problem, especially in a country such as the United States, with its relatively high incidence of HIV and its correctly demanding safety standards. So either a method needs to be devised for shielding the waveguide from direct contact with the blood, or waveguides must be disposable, or one waveguide should be assigned to each patient for his/her exclusive use.

How many BI treatments can be safely given?

Various Russian and German authors mention 15-20 treatments or even as many as 30. As long as there is an interval that will permit recuperation or replacement of sensitive cells--perhaps two months between treatments--BI can probably be used indefinitely in the case of a chronic disease that needs to be suppressed at regular intervals. The danger of inducing a cancer at such low levels of radiation is minimal.

---

[35] Ibid., pp. 140-3

[36] A.P. Vasilyev et al., "Changes in Blood and Red Cell Membrane Lipids in IHD Patients under the Influence of Quantum Therapy [Russian]," Terapevticheskii Arkhiv 68 (1996), No. 12, pp. 47-50

[37] V.A. Ovsiannikov, "Analysis of the Low-Energy Laser Treatment of Some Cancers and Infectious Diseases," Journal of Clinical Laser Medicine & Surgery 15 (1997), No. 1, pp. 39-44

But isn't there a possibility that the irradiation could cause long-term damage?

It is true that no long-term studies of BI's effects have been done. However, BI is much lower in intensity and far less concentrated on a specific target than were the x-ray treatments of the 1930s and 1940s that led to cancers decades later. The relatively rapid turnover of the blood cell population also reduces the impact of BI. In contrast to x-rays, little of the UV radiation from the Russian UBI device is ionizing; in UV-A devices, none of it is. None of the currently used LBI devices emits ionizing radiation.

Logic suggests that BI has a prophylactic action against cancer. An East German study of mutagenicity in chromosomes before and after 6 sessions of UBI found that in fact chromosomal aberrations had diminished in number, leading to the hypothesis that UBI may actually stimulate DNA repair.[38] There is also not a shred of evidence that properly dosed BI damages any specific organ or tissue of the body such as lymph nodes other than the minor damage that it does to the membranes of many red and white blood cells. Even there, in applications such as the treatment of preeclampsia, there is evidence that BI stabilizes membranes against lipid peroxidation.[39]

If BI is so safe and effective, why is it so little known outside of Russia and Ukraine?

1) Medicine in Western and technologically advanced East Asian countries has gone down the path of molecular biology. Physicians and researchers trained in biochemistry (and often with scant knowledge of physics) sometimes look askance at biophysical approaches. 2) Statements regarding BI can easily be associated with the myriad specious claims of wonderfully curative devices by enthusiasts and charlatans. 3) The present association of BI with things Russian can hurt it in the eyes of those who are aware of the financial and technical weaknesses of the Russian medical system. 4) The general low prestige of Soviet and East German communist systems as well as the lingering effects of Cold War propaganda against them have led to a tendency to

[38] Frick (1989), pp. 55-6

[39] A.S. Bednarskii et al., "The Use of Intravascular LBI in Combination Therapy of Preeclampsia [Russian]," Akusherstvo i Ginekologiia (1995), No. 6, pp. 18-22

belittle the genuine contributions of their scientists. 5) The lack of Russian- and German-language skills among Anglo-Saxon and East Asian researchers leads to a mentality in which it is hard for some to accept that there could be a cutting-edge therapy like LBI on which almost none of the scientific literature is in English. 6) Many physicians have surprisingly little knowledge of the real history of their own specialties; they know the textbook history and the medical literature of the past 25 years, neither of which includes BI. 7) In turn, this leads to an NIH (Not Invented Here) syndrome--the irony being that BI *was* invented here, and the Russians were Johnnies-come-lately to it. 8) Since the mid-1950s, the few American practitioners of UBI have chosen to treat patients quietly rather than do battle with state medical boards. The effectiveness of UBI ensures that they do a steady, lucrative business in patients who prefer their services to those of colleagues. 9) Organizing clinical trials to validate BI would require considerable effort and financial resources. 10) Finally, the difficulty of discovering the underlying mechanism of action of BI long deprived its advocates of a valuable weapon.

But surely American advocates of alternative medicine would have displayed interest in BI if it were effective?

Knowledge of Russian, German, physics, and the real history of medicine is not exactly their strong point either. In addition, BI never was part of some broader approach to medicine; its links with therapeutic UV irradiation of the skin were always weak. So, unlike a Chinese herbal remedy, for instance, BI has no long tradition and no large, permanent constituency to back it. Thus one can consult many handbooks of alternative medicine that discuss the most arcane and dubious therapies without finding a single reference to BI. When BI is mentioned, it is almost invariably without any true comprehension. BI slipped through the cracks between standard and alternative medicine.

However, there are two valuable recent books on BI by American physicians,[40] and certain American physicians are expert in its use.

Don't the National Institutes of Health have the responsibility for investigating innovative therapies?

---

[40] Miley (1997) and William C. Douglass, *Into the Light*. 2nd ed., Atlanta: Second Opinion, 1996

Yes, but NIH has a track record of dismissing promising, original approaches that eventually turn out to be highly beneficial. Its lapses in regard to Magnetic Resonance Imaging and the homocysteine theory of heart disease are cases in point. NIH has known about BI for years, but it has refused to put a penny into seriously investigating it. In the light of the formidable evidence regarding the effectiveness and safety of BI, dismissive responses from NIH should be considered more a matter of curiosity than of consequence.

Also, it is worth noting that BI can no longer be termed "innovative." It has been around for 70 years, has been exhaustively studied, and is well-characterized.

In what circumstances can BI devices be used?

The portability of the devices makes them suitable for use in a variety of settings. A nurse can bring a UBI device to the homes of bedridden patients, with the added benefit of ensuring compliance. In developing countries the UBI device can be used by medics in remote villages that have electricity under the long-distance supervision of a centrally located physician. Or a traveling physician can take a low-intensity laser or UBI device on his/her rounds. UBI units can also be deployed in battery fashion in a large clinic under the supervision of a single nurse (with an aide to assist in setup and cleanup), though it is advisable to have only two patients per nurse at any time and one-on-one is better to monitor the patient and insure that the blood does not coagulate.

So BI could be of value in the Third World?

The modest cost, simplicity of use, and versatility of Blood Irradiation peculiarly suit it to the needs of Third World countries. If the fragmentary findings of BI's effectiveness against chronic hepatitis can be validated, for instance, BI could significantly reduce the incidence of chronic hepatitis in Africa.

Does concomitant administration of other therapies influence BI's effectiveness?

Some practitioners warm up an inflamed joint prior to administering BI to enhance specificity, though it is not clear if this procedure has any effect. Magnet therapy has been used in conjunction with LBI in

Russia. Concomitant administration of glucocorticoids, non-steroidal anti-inflammatory drugs, and Vitamins C and E diminishes the effect of BI. Logic suggests that antimetabolites like methotrexate and various nucleoside analogues would also reduce BI's effectiveness. Sulfa drugs and other photoactive drugs should not be administered directly after BI in normal (e.g., high for the circumstances) doses; their excessive activation can be dangerous.

The early American practitioners found that UBI was most effective as a monotherapy. In a retrospective study of 9 cases of typhoid fever, the 3 patients treated with UBI alone recuperated more rapidly than 3 treated first with sulfa drugs, then with UBI. Of 3 treated with sulfa drugs alone, 1 died and 2 had very long recuperations.[41]

Curiously, in a 1943 study of bronchial asthma, there is evidence that the researchers were systematically understating their results, in part because of excessively strict criteria. Of the 9 patients under the age of 20, the 8 who remained in the study were consistently given scores that were below the written characterizations of their end status.[42] This "self-abnegating scoring" is characteristic of a situation in which researchers slant their results in a negative direction in order to avoid hostile criticism that they are exaggerating them.

The early American proponents of UBI appear to have deliberately understated their case. They evidently thought that it was so compelling that their colleagues would eventually grasp the point, and they wished above all to avoid any sensational claims that would allow critics to lump them with the many kooks who have always been attracted to electromedical devices. At one point, George Barger, M.D. even suppressed an article on UBI by the father of a successfully treated child that was scheduled to appear in a popular magazine.

One weakness of the Russian clinical trials of BI is that standard drug regimes are often continued even during BI treatment. Ironically, far from being indiscriminate in their employment of BI ("zapping the patients", as the crude Western image would have it), Russian physicians have generally been excessively cautious. Their unusually sensitive applications of BI (e.g., against fetal distress) have been the outcome of step-by-step, painstakingly observed and tested advances

---

[41] E.W. Rebbeck and H.T. Lewis, "The Use of Ultraviolet Blood Irradiation in Typhoid Fever," Review of Gastroenterology 16 (1949), pp. 640-9

[42] G.P. Miley, R.E. Seidel, and J.A. Christensen, "Intractable Bronchial Asthma, Preliminary Report of Results Observed in 80 Cases," Archives of Physical Therapy 27 (1943), pp. 533-42

against less sensitive conditions. This means, however, that BI is probably even more effective than appears in the results of these trials if used as a stand-alone treatment. In addition, most recent Russian trial data pertain to LBI, which is demonstrably somewhat less effective than UBI.

In East Germany, out of a concern over possible increased mutagenicity resulting from BI, practitioners deliberately steered clear of treating younger patients, even though they acknowledged that there was no evidence to support this concern and their own studies showed it to be unfounded. As a result, their clinical results--though excellent--were derived from a patient population that was much older (53.7 years) than average and thus considerably less likely to respond as favorably as younger patients would to BI; and they treated sexually transmitted diseases and other major indications for BI only rarely. As for the earlier German practitioners, they ordinarily used intramuscular injection of extremely low doses of irradiated blood, a less effective approach than the Knott Technique.

In other words, it appears that--far from exaggerating--American, Russian, and German studies of BI systematically understate its potential effectiveness.

Has any practitioner ever used BI in an optimal manner?

Yes, Robert C. Olney, M.D. As a senior surgeon based in Lincoln, Nebraska, Olney had the intuitive understanding to push UBI to its limits because he saw how low its toxicity was. Over the course of 30 years he treated many thousands of patients for a wide range of serious indications. He published a series of highly interesting studies, e.g., of the use of BI in pelvic inflammatory disease where he achieved a ratio of 80 percent totally successful outcomes in severe, refractory cases. He resorted to surgery only in the 20 percent of cases in which there was a clear cyst tumor, abscess, or other obstruction.[43] Olney repeatedly reported results superior to those of his colleagues, even in difficult indications.

What can explain this record? 1) It is possible that he was exaggerating. 2) The many wavelengths of the Knott Hemo-Irradiator made it more effective than other devices. 3) Olney used UBI as a monotherapy except in the case of post-heart attack conditions. 4) He

---

[43] R.C. Olney, "Ultraviolet Blood Irradiation Treatment of Pelvic Cellulitis, Knott Method," American Journal of Surgery 84 (1947), No. 4, pp. 440-3

would administer as many treatments as the patient's condition required, secure in the knowledge that as long as there was a diseased state in the body, BI would do no harm. There is a good deal of evidence that Olney was not exaggerating at all and in fact was an outstanding physician.

Are there any special modes in which BI is being administered?

Russian physicians have irradiated marrow cells to treat osteomyelitis, cerebral spinal fluid to treat multiple sclerosis, and portal blood to treat hepatitis. Ukrainian physicians are experimenting with pulsed LBI timed to the patient's heartbeat in order to optimize the effects of a given dosage and, with the help of computers, take a step toward the goal of real-time monitoring of LBI. In Odessa they are experimenting with the use of noncoherent light for BI.[44] The Ukrainians also now have developed a laser device that irradiates blood in an extracorporeal mode.

An occasional practice is to use irradiated donor blood, though it is not clear what advantage this might confer over autologous transfusion of irradiated blood. However, in the literature of BI there are no reports of serum hepatitis following transfusion of UV-irradiated donor blood. This led Russian researchers to conduct a study of the irradiation of 31 samples of donor blood containing antigen to hepatitis. They found that UV-C irradiation removed any trace of antigen from 67 percent of the 15 samples irradiated with it, and UV-A had a score of 50 percent of 16 samples. They recommended that this method be used to reduce the risk of transmission of serum hepatitis.[45]

This finding suggests that BI of all donor blood might reduce the transmission of pathogens in two ways: via direct irradiation and via the *in vivo* action of BI's secondary radiation. This approach could be especially useful in countries with lower quality control of donor blood.

Are there any counterindications to the use of BI?

The original Soviet regulatory documentation for the UBI device listed the following counterindications: 1) difficult cases of cardiac insufficiency of the left ventricle failure type; 2) within three weeks of

---

[44] See Ukrainian Center for Laser Medicine. *The Use of Lasers in Biology and Medicine* [Russian and Ukrainian]. Kiev, 1996

[45] A.V. Marchenko et al., "The Possibility of Using UV-Irradiated Blood in the Prophylaxis of Serum Hepatitis during Hemotransfusion in Surgery [Russian]," Vestnik Khirurgii Imeni I.I. Grekova (1990), No. 2, pp. 87-89

myocardial infarction; 3) acute cerebral insufficiency and acute disturbances of cerebral circulation; 4) gastroduodenal hemorrhage; 5) photodermatitis; and 6) hypoglycemia.

Subsequent clinical studies in Russia and Ukraine make it clear that BI is much safer and more effective in heart disease than these regulations would suggest.[46] Many clinical trials of BI in neurological and psychiatric disorders suggest that it is safe in acute cerebral disturbances as well, though this question deserves further study. The small risk of hypoglycemic shock can be avoided by feeding an at-risk diabetic patient carbohydrates just before or after BI treatment.

Common sense dictates that, as with any therapeutic intervention, these devices be used sparingly on small children. They have, however, been used in Russia in careful studies with great effectiveness and safety to correct fetal conditions hard to treat with drugs as well as infections, hypoxia, and slow growth of newborns.[47] Lastly, they should not be employed for trivial purposes.

What are the known side effects of BI?

1) Flushing in some cases. 2) Creation of a small amount of ozone. 3) Destruction of some immune cells, depending on the dose and the condition. 4) In cases of disseminated infection, the rapid destruction of high numbers of infectious microorganisms can temporarily create toxic symptoms that subside as the microorganisms are cleared from the blood. 5) In 50 percent of bronchial asthma patients, there is a flare-up of symptoms following the first treatment with BI; similar flare-ups can occur in rheumatoid arthritis. Subsequent treatments are uneventful.

While some experienced practitioners consider BI to be without any significant damaging side effects whatsoever, one well-informed German source[48] reported side effects in 15.3 percent of cases (84 of 550), including hypoglycemic shock (4 cases, probably diabetics or others with a tendency toward hypoglycemia), allergy (10), tiredness (7), fever (7), inflammatory responses in tooth root granulomas (5), gastritis (4, one case of which required cessation of BI therapy), and

---

[46] Yu. N. Sirenko, I.E. Malinovskaya, and S.S. Krasnitskii, "On the Treatment of Patients with Severe Coronary Insufficiency with Ultraviolet Blood Irradiation [Russian]," Vrachebnoe Delo (1990), No. 10, pp. 9-11

[47] A.I. Matsuyev et al., "The Use of Ultraviolet Blood Irradiation in Obstetrical-Gynecological Practice [Russian]," Akusherstvo i Ginekologiia (1990), No. 8, pp. 6-10

[48] Frick (1989), pp. 54-55

exacerbation of asthma (2, in one case requiring cessation of BI therapy). He admitted that his list included various phenomena that may not be connected with BI at all and that many of the reactions were trivial. He regularly administered up to 10 treatments of UBI, and sometimes more, at frequent intervals. Lastly, his patients were 53.7 years old on average, so they were a good deal more likely to report side effects than a more resilient youthful population.

A Russian study of 2,380 sessions of UBI revealed that 1.3 percent of patients had minor complications--hematomas at the IV site, coagulation in the tubing, shivering, dizziness, and nosebleeds. In addition, one had hypoglycemia, one had bronchospasms characteristic of her reaction to other treatments, and one had a nettle rash (urticaria).[49]

In over 10,000 treatments with UBI (eventually he would perform over 16,000), Miley noted six citrate reactions typical of IV procedures. They included headache, chill, temporary fever up to 102 degrees F, and moderate gastrocnemius spasm. Miley noted that the administration of sulfa drugs *following* UBI, which occurred against his repeated warnings, could be fatal in weakened patients. He recommended that sulfa drugs, iodides, and digitalis not be administered for 4-7 days following UBI. Presumably this applies to other photosensitive drugs as well.

Miley also observed mental lassitude and sleepiness in some cases as well as coagulation problems owing to faulty technique.

Lastly, he noted that in several hundreds of thousands of UBI treatments, there was one instance reported of a serious reaction in an acutely septic patient characterized by violent chills, fever, and severe erythema followed by death--possibly a result of a reaction to fibrin owing to improper cleaning of the device.[50]

The Russian, German, and American findings suggest something very plausible: that BI actually has a profile of damage to vulnerable sets of body cells that closely parallels that of many chemotherapies--in particular, the nucleoside analogues. The difference is that BI's greater specificity gives it a considerably higher therapeutic ceiling than competing chemotherapies (or herbal remedies, for that matter), so such damage only occurs with a relatively larger dose of BI.

---

[49] A.B. Marochkov, V.A. Doronin, and N.N. Kravtsov, "Complications in the Ultraviolet Irradiation of the Blood [Russian]," Anesteziologiia i Reanimatologiia (1990), No. 4, pp. 55-56

[50] Miley (1997), pp. 21-22

Have the UBI device or LBI lasers been tested in controlled clinical trials?

Both UBI and LBI have been tested extensively in clinical trials, particularly in Russia and Ukraine. In recent years these trials have been employing increasingly strict protocols, including controls and sophisticated statistical analysis, though double-blinding and multicenter trials are still not the norm and it is not clear in many cases how strict the randomization is. Another difficulty in assessing BI's effects arises from the tendency to employ it as part of combination therapy rather than as a stand-alone treatment. And often there is no report of long-term follow-up.

Several trials and studies with historical controls were carried out in the US, but none since around 1960. Much of the early American reporting on UBI consists of series of case reports. In general, the East German, Russian, and Ukrainian results have a higher validity than earlier American ones. Their laboratory and clinical studies have been more extensive as well and are based on a much more sophisticated understanding of immunology and general medicine than was available 50 years ago. In addition, they have not been subject to the insidious commercial forces that shape and sometimes corrupt the clinical trial process in Western countries.

What is the relationship between UBI and photopheresis?

In the 1980s Yale University researchers independently developed a method of irradiating the blood that is termed "photopheresis" or Extracorporeal Photochemotherapy.[51] Photopheresis employs drugs, filters, and separation of the white blood cells from the red blood cells in the plasma. This treatment costs $2,000, requires sophisticated equipment, and takes many hours. Photopheresis uses a low-intensity fluorescent source of UV-A while the Russian UBI device employs a high-intensity mercury-quartz source of UV-B and UV-C. Many medical centers now use photopheresis.

---

[51] R. Edelson, "Light-activated Drugs," Scientific American 259 (August 1988), No. 2, pp. 68-75. This article did not contain any mention of UBI or the history of its use although the author had cited the original 1928 UBI device patent in his own patent application.

In effect, photopheresis is a combination of UBI and chemotherapy in which the secondary radiation triggers the photoactive drug previously taken up by the target cells. Thus to achieve the same effect, photopheresis uses less irradiation and more chemotherapy than UBI. The substances used are generally psoralens, which occur in nature but are used in chemotherapeutic concentrations that can have more toxic effects than other forms of BI (Figure 1 shows how the 3-ring psoralen is intercalated into DNA, thereby replacing the proper linkages between adenine and thymine).[52] The two therapies appear to have roughly the same effectiveness, with UBI presumably having an edge with equal doses of "medicine" (i.e., of toxicity) because of its higher specificity. Photopheresis is probably effective for most of the indications UBI is effective for, and the reverse is presumably also true.

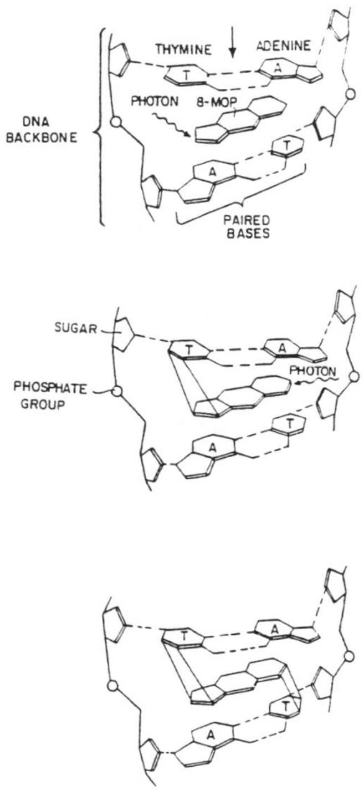

Figure 1. DNA Intercalation in Photopheresis

---

[52] R. Edelson, "Photopheresis: a Clinically Relevant Immunobiologic Response Modifier," Annals, New York Academy of Sciences (1991), Dec. 30, p. 636

Photopheresis has these comparative advantages: it is approved by the FDA for the treatment of cutaneous T-cell lymphoma; it is currently in clinical trials for other indications; hundreds of photobiologists have studied it; there are many recent English-language publications on it; and it is available in many medical centers.

UBI has the comparative advantages that both the device and the treatment are much less expensive; the duration of the treatment is briefer; the UBI device can be used by any individual with basic medical training; the device is more portable; UBI has been used on a wider range of indications; and UBI's activation of red blood cells temporarily transforms them into a dynamic component of the immune system.

Two additional considerations are that some step in the procedure of photopheresis (e.g., centrifugation that permits the UV irradiation to concentrate on lymphocytes) might confer an advantage on it; and, conversely, the apparent exceptionally high specificity of UBI may make it "cleaner" than photopheresis, which relies on chemotherapy and has regular side effects. A comparative trial of photopheresis and UBI could shed light on both of them.

An important implication of the FDA approval of photopheresis is that the FDA thereby accepted the principle that a therapeutic use of BI could be both safe and effective.

How does LBI fit in?

Everything that is said about UBI's effects appears to apply to LBI as well, though there are subtle differences in their patterns of action and effectiveness.

Isn't it hard to believe that all of the secondary radiation from BI is channeled directly to the problem spot without any ending up in the wrong places?

Yes. In fact, it is clear that some of the secondary radiation directly affects the neurons that activate the autonomic nervous system (hence the flushing of the skin), and some of it stimulates the entire immune system. In addition, some secondary radiation is dispersed around the body. All erythrocyte membranes appear to be altered somewhat, for instance.

One explanation for the lack of observable side effects would be this:

Divide the cells of the body and any infectious microorganisms in it into three categories of energy-demanders: A--high; B--moderate; and C--low. Into Category A would fit active infectious agents and activated immune cells. These would absorb an inordinately large share of the available blood glucose and, in parallel, of the secondary radiation of BI. The cells in Category B--those cells in the stomach, mouth, brain, and elsewhere with somewhat higher metabolism than ordinary body cells and therefore the ones most often damaged by chemotherapy--would absorb only a small amount of secondary radiation because the blood's energy gradient would direct the main pulse of it toward the infectious microorganisms or activated immune cells (A). Meanwhile, the overwhelming majority of body cells would belong to Category C. Billions of them each would absorb an amount of secondary radiation equivalent, perhaps, to a few stray biophotons.

All three types of radiation--the initial irradiation with UBI (LBI obviously differs somewhat but not in essence), the secondary radiation from the irradiated blood, and the ultraweak radiation emitted by cells in the form of biophotons--are in the ultraviolet band of the spectrum, so in this sense BI is "natural" not only because of its similarity to sunlight but also because of its similarity to the ultraweak radiation of the body cells. In a word, BI is right at home in the microambience of the body in a way that no chemotherapy can ever be. In contrast to the barriers the cells might set up to block out chemotherapies perceived as somehow incorrect, they would readily accept their minuscule portion of BI's secondary radiation as a form of natural energy. Their cellular mechanisms could readily repair any damage such a tiny amount of secondary radiation might do; it is very unlikely that a few stray photons do much damage anyway. In turn, the cells' readiness to absorb some of the secondary radiation would reduce the amount that might otherwise end up in the cells in Category B. In effect, the billions of cells in Category C act as an enormous ecological catchment basin.

The philosophically inclined might view BI as a link between macrocosm and microcosm.

Then what happened to Knott's poor dogs?

Once the secondary radiation from BI had destroyed all of the infectious microorganisms (Category A), the blood then channeled the still incoming radiation in the direction of the cells in Category B, thereby overwhelming them in a manner similar to what would have occurred with massive overdoses of various chemotherapies. The difference between such chemotherapies and BI in the period before the total destruction of infectious microorganisms of Category A is that the drugs' deviation from the ideal forms of energy (glucose, ATP, BI's secondary radiation) causes the blood's energy gradient to channel them less precisely and so they damage Category B cells from the very outset, especially if they are also rejected by the resting cells of Category C.

Implicit in this model is the assumption that the blood has an ecology in which the top predators (at first Category A, then Category B once Category A is destroyed) can lay claim to an exceptionally high proportion of the surplus energy in the blood, whether in the form of glucose or radiation.[53] Another assumption is that pharmacokinetics (and within it the primordial question of specificity) is more important in explaining the action of certain drugs than is pharmacodynamics. How the concentrated impact of the secondary radiation of BI, once absorbed, destroys an activated virus or T-cell is an interesting scientific question; but it is less important from the standpoint of the balance between therapeutic effect and damaging side effects than the question of how the radiation is concentrated in that cell and not elsewhere in the first place.

Is there an underlying law of pharmacology that explains the kind of specificity observable in BI?

Yes, the Law of Energy Specificity: The more closely a substance resembles forms of energy and the greater the metabolic activity of cells and infectious agents, the higher the specificity with which the substance is channeled to and absorbed by them.

Why is specificity so important?

Because to say that a treatment has high specificity is tantamount to saying that it is effective and safe.

But what about those cases in which BI therapy is not effective?

In many of them, the explanation might be that the disorder has progressed to a stage at which it has disrupted the normal functioning of the body's energy mechanism.

---

[53] See Dillon (1994), p. 42.

How can such small doses of irradiation have such powerful effects?

Blood cells contain various forms of energy and are highly reactive. The initial irradiation triggers a release of energy from them in the form of secondary radiation. The amount of secondary radiation emitted might well exceed (or be less than) the level of irradiation initially received. Even the relatively low amount of ambient radiation absorbed during the period that blood is removed from the body for purposes other than BI is known to have a beneficial effect upon retransfusion. This effect is one cause of the overall benefit of UBI, though not of LBI.

How can the blood emit more energy than it receives?

The blood can be defined as a rather well-organized, energy-bearing fluid with the properties of a resonant cavity. This characteristic arises from the sum of the responses of the blood cells, whose cellular walls make them resonant cavities, but the biophysical properties of the cellular field also play a role. In other words, the blood operates as a distinct system that, through a series of branching chain reactions, reacts in a nonlinear manner in response to inputs of energy.[54] BI "turns the cells on" rather than simply "charging" them, though there is an element of charging.

An important implication of this is that the parameters for the intensity of irradiation of the blood in BI are quite wide because the irradiated blood sample will proceed to react according to its own principles. Thus the secondary radiation it emits might bear little relationship to the precise level of irradiation it receives. In turn, this phenomenon may explain why considerable variation in treatment intensity, duration and number of sessions, source of irradiation, and intervals often does not seem significantly to affect the therapeutic result. It is true, however, that at times the dosages employed in clinical trials seem inadequate to the task.

The relationship between level of irradiation and therapeutic outcome is a priority subject for BI research. It is possible that the number of times the treatment is repeated is more important than the

---

[54] See V.L. Voeikov, C.N. Novikov, and N.I. Siuch, "Alterations in Luminol-enhanced Chemiluminescence from Nondiluted Whole Blood in the Course of Low-level Laser Therapy of Angina Pectoris Patients," SPIE Proceedings, Vol. 2895 (1997) and the references therein.

level of the dose; the signalling effect of the irradiation may be more significant than the charging one. Likewise, repeating the treatment on a daily basis may be more effective than doing so at greater intervals, either because in this way BI mimics the periodicity of solar irradiation or because the initial activation of the cells does not fully subside and thus is taken to higher levels on subsequent days of treatment.

Is BI uniformly effective against overwhelming infections and fulminating autoimmune conditions?

No. In some cases, the patient's condition has deteriorated beyond the help of BI. In others, BI simply is not effective. Still, no other therapeutic intervention appears to work nearly as well as BI in such difficult indications.

Is BI equally effective against all kinds of microorganisms?

BI's action against fungal and protozoan diseases is not well characterized. The higher the proportion and activation potential of DNA in a microorganism, the greater is its vulnerability to radiation (as well as to energy-mimicking drugs). Viruses are thus the easiest target for BI's secondary radiation. Then come various bacteria, fungi, and protozoa in a ranking that would become evident in response to therapy but difficult to validate according to relative DNA proportion and activation potential. Of course, BI also activates the entire immune system, so it is not easy to sort out its exact workings.

What explains BI's variation in effectiveness in a given disease?

The most important cause seems to be that, like other interventions, BI is more effective during the early stages than it is after irreversible damage has occurred. Failure to take this into account has led to considerable confusion and disappointment. BI probably retains its therapeutic effect farther into a disease's course than competing therapies, but one should not expect miracles.

A second cause of variation in BI's effectiveness is that frequently BI is not provided in sufficient dosage or number of sessions. A third, less significant cause might be variation in the output of the UV source in UBI. This is not a problem in LBI.

Aside from its therapeutic applications, does BI have other uses?

In certain cases of difficult differential diagnosis, a patient's response to BI might be of diagnostic significance. More generally, the responses of various syndromes to BI may help to identify their underlying mechanisms of action. In a similar fashion, the clinical use of BI can clarify the mechanism of action of other therapies. For instance, LBI's effectiveness against schizophrenia with depressive syndrome is clearly attributable to its ability to unblock microcirculation in the brain by destroying activated white cells and platelets.

This finding may provide the answer to the old riddle of the mechanism of action of Electroconvulsive Therapy: ECT may have exactly the same effect on white blood cells and platelets. In turn, this suggests that BI is both safer and more effective than ECT and that it can successfully treat certain patients resistant to ECT. This conclusion can be drawn from a Russian study, though the researchers do not draw it.[55] Some of the patients in this trial had previously failed ECT, but the researchers do not make it clear whether these patients were among the many subsequently successfully treated with LBI or not. Another conclusion that might follow from the use of BI in this and other neurological indications is that various disorders of which the mechanism has hitherto been disputed may now be shown to be caused by microcirculatory blockages.

In a broader sense, BI's role as an antagonist/substitute/overrider of glucose and ATP endows it with unusual scientific interest in terms of affording insight into the body's energy metabolism. The dynamics of the process by which the irradiated cells transmit secondary radiation as well as the ways in which this radiation is absorbed throughout the body, for instance, can shed light on fundamental cellular mechanisms and phenomena such as biophotonic radiation.

So BI is of scientific interest?

Definitely. Among other things, there is evidence that in BI certain phenomena occur that are not accounted for by the standard laws of photochemistry. A study of BI's action might also shed light on the

[55] I.I. Kutko, V.V. Pavlenko, and E.G. Voronkov, "The Use of Intravascular Laser Blood Irradiation in the Treatment of Forms of Schizophrenia Resistant to Therapy [Russian]," Zhurnal Nevropatologii i Psikhiatrii Imeni S.S. Korsakova 92 (1992), No. 4, pp. 53-56

nature of the "communications phenomenon" whereby cells are said to signal to each other by means of patterns of biophotons. Likewise, demonstrably safe and effective doses for LBI in clinical practice can differ considerably from prescribed norms.

BI's mechanisms of action may help explain seemingly parallel mechanisms in magnet therapy.

Are there any other uses of BI?

BI can be used to oxygenate the blood of athletes, though this is a dubious application.

Still, BI sounds a bit like certain therapies that are promoted by quacks, doesn't it?

Yes.

So it is acceptable to be skeptical?

Perhaps initially, but it is a mistake not to investigate farther. One helpful approach is to keep in mind the magisterial dictum of Paracelsus: "All things are poison, and nothing is without poison. It is the dose alone that makes a thing not a poison." From this perspective, even the most appealing and non-intrusive form of natural medicine can be a poison if it is employed to excess or in improper circumstances. One could do serious damage to a patient with a massive overdose of BI, just as one could with a massive overdose of aspirin, herbs, or any other drug or natural remedy on the market. But with the right dose of BI, one can bring back to good health a patient with one foot in the grave.[56] In fact, the exceptional specificity of BI gives it a very wide range of therapeutic benefit, making it potentially safer than all competing therapies.

In addition to the very clear, consistent pattern of effectiveness reported in studies by scores of researchers in different countries at different times, there is striking internal evidence that shows how trustworthy the sources and information are. For instance, the forebearance of Knott and his medical collaborators in waiting five years after the initial highly successful treatment in 1928 in order to observe

---

[56] R.C. Olney, "Ultraviolet Blood Irradiation, Knott Method, in Biliary Disease," American Journal of Surgery 72 (1946), p. 235

the first patient before treating a second one was a remarkable example of scientific probity. These were serious, ethical scientists.

Another telling piece of internal evidence is the consistency of the results of the Vladivostok bronchial asthma trials. In four trials in a row involving many hundreds of patients, UBI repeatedly outperformed LBI in exactly the same way. In a fifth, on patients with chronic bronchitis, the results of UBI and LBI were virtually equivalent; but the researchers had deliberately assigned the more severe cases to the UBI group, so that this result confirmed those of the bronchial asthma trials. In other words, the finding that UBI is more effective than LBI is very robust, the more so because the researchers made it clear at the outset that they would much prefer to use LBI exclusively for several reasons.

To voice skepticism about findings of such power is a patent mark of bias. Once one accepts that one version of a therapy is more effective than other, however, it would be somewhat contradictory to turn around and voice skepticism about the therapy's effectiveness in general.

Although Knott's initial results in the dog experiments should have made it clear that BI is no mere placebo, certain critics persist in attributing its effect to psychological causes. American, German, and Russian researchers have, however, repeatedly studied this question in clinical trials. They have consistently found that a placebo-like effect occurs frequently but that even when it does BI's physiological action indisputably surpasses placebo.

In circumstances where it is easy for critics to indulge in unbridled skepticism (the Russians make a tempting target for endless aspersions as does the superficial resemblance BI bears to therapies promoted by quacks), it is essential to put the critics themselves at risk. One difficulty in doing so is that many of them appear to have no incentive whatsoever to avoid a Type 2 error, a false negative conclusion that would lead them to dismiss a very promising therapy.

How can one best investigate further?

The Bibliography contains selected pertinent articles and books. Another source is the US National Library of Medicine's Grateful Med citations, available on the Internet, to the latest clinical results obtained by Russian and Ukrainian physicians. The track record of photopheresis can offer further insight.

One approach might be to consider the lack of effectiveness of other therapies and the collateral damage they cause, then to give thought to whether BI might be a more fitting treatment for the indications in question. No therapy should be judged in a vacuum or according to impossible standards of perfection. Only by comparing various therapies according to several practical criteria, including cost, can one make optimal judgments.

Thus the question "Does BI work?" is not a useful one because it fails to place the therapy into a context. In a sense, all therapies "work." One could speak of a Principle of Therapeutic Correspondence: Every source of energy has a corresponding therapeutic range. The proper questions with UBI, LBI, or any other medicinal or biophysical therapy are: "What is its therapeutic range? What are the circumstances in which it is appropriate to use, and what effects does it obtain in those circumstances? How does it compare to other therapies? What are the counter-indications?"

Similarly, the question "Is it safe?" is not helpful. It can lead to a bottomless pit of doubt whereby every piece of evidence of the safe application of BI is met with the further question: "But isn't it possible that BI causes some hidden, systematic damage of which we are not yet aware?" That approach is ultimately paranoid. The correct scientific questions are: "What are the level and pattern of BI's toxicity? How do they compare with those of competing therapies?" In fact, these are simply other ways of framing the above questions on effectiveness.

Are any practitioners in the United States currently using the UBI device?

Some 20 practitioners in the United States are using it. Thousands of Russian and Ukrainian physicians regularly use UBI and LBI, as do some of their German colleagues.

Other than the low cost per treatment, are there any other pertinent cost-related considerations?

Yes. Russian researchers repeatedly report significant reductions in the length and frequency of hospital stays because BI is more effective than competing chemotherapies in many indications. In addition, many patients with debilitating diseases that have made them invalids are able to return to work following BI. Timely intervention with BI can also save on the expense of operations such as amputations of diseased legs in diabetes.

Is there any danger that the UBI device will transmit disease from one patient to the next?

The needles and tubing are disposable. The quartz irradiation chamber (cuvette) is cleaned and sterilized with a technique that destroys all contamination. Still, to be safe, it is best to provide one cuvette to each patient for his/her exclusive use or to use disposable cuvettes. With LBI, needles should be disposed and waveguides not shared among patients. One danger with both UBI and LBI in poor countries is that some practitioners will be tempted to reuse disposable cuvettes and waveguides.

There must be some *drawbacks to BI?*

Careful study has revealed two drawbacks.

First, as with all drugs (for although BI is irradiation, it has many of the characteristics of a bloodborne drug), it is necessary to provide the right dosage. Too little will not have effect; too much could do damage. This is not too difficult with LBI, but with UBI the problem is one of establishing objective measures for treatment. Thus far no researcher or practitioner has developed a way to measure the optimal wavelength, duration, or intensity of irradiation, or the optimal amount of blood irradiated. For this reason, Knott and his medical colleagues preferred to use modest doses and to repeat the treatment after an interval during which they could observe the patient. They established guidelines for dosage and intervals, but a considerable latitude remains for the practitioner's judgment.

This characteristic of UBI makes it almost impossible to define what an optimal dose might be. It has, however, the benefit of obliging practitioners to individualize therapy and thus to avoid the problems that can arise from excessive adherence to protocols. The lack of observable side effects from UBI also suggests that even in the case of overshooting somewhat during an individual session of UBI, the potential damage is vanishingly small. Such overshooting can occur in chemotherapy administered according to protocol as well and can do considerable damage.

In other words, an excessive pursuit of precision is the enemy of efficacy in BI therapy. In their understandable desire to attain precision and ease of use, many Russian practitioners have abandoned UBI in favor of LBI. But now it is clear that in doing so they chose the second best approach (though LBI is certainly very good medicine). BI requires the practitioner to recognize that medicine calls for more than mere technical skill; it has an inescapable element of judgment and art to it as well.

A second drawback of BI is that in certain indications, for reasons that are not entirely clear, patients' responses to BI are not as uniform as might be expected. For instance, a rigorous Ukrainian clinical trial of LBI in rheumatoid arthritis found that treatment actually exacerbated the worst cases, presumably where irreparable joint damage had already occurred, perhaps by provoking the immune system to react even more strongly. In contrast, LBI was consistently very beneficial for patients with mild and moderate arthritis of shorter duration.[57] A careful comparative study might reveal that the limits of BI in severe cases of rheumatoid arthritis closely parallel the known limits of methotrexate, the current standard treatment.

Any other problems?

An obvious initial limitation to the use of BI is the patchy, anecdotal evidentiary basis for its application in specific indications.

Does BI have any advantages over competing chemotherapies?

BI therapy has many advantages over chemotherapies:

- BI's higher specificity gives it greater effectiveness;
- BI is less likely to induce drug resistance in microorganisms. The secondary radiation of BI therapy has more uniformly destructive effects on microorganisms than do chemotherapies, which some microorganisms may reject or ingest and develop resistance to. For a virus to develop resistance to radiation, it would have to evolve into something quite different: not impossible, but certainly not an easy task. Certain parasites, however, might not be vulnerable to therapeutic doses of BI not because they have developed resistance

---

[57] Zverova et al. (1994), pp. 29-32

but because they were resistant to start with, perhaps because of an ability to freeze their metabolic processes;

- because BI therapy is so versatile, a physician can become a master at its use for many purposes, thereby saving on the time required to learn the details of many drugs for specific indications as well as avoiding the possibility of making an error in using a new drug;
- the problem of compliance with BI therapy is much smaller than with chemotherapies, in turn lessening the likelihood of the development of resistance and spread of infectious diseases;
- except in odd circumstances, accidental or deliberate overdoses by patients are impossible with BI. An experienced practitioner is in control of the device and its use at all times;
- unlike antibiotics, BI does not destroy benign flora, does not depress the immune defenses of the body, only occasionally gives rise to allergic reactions, and has no toxic effects on specific organs;
- in acute infectious diseases of unknown etiology, BI can immediately be employed, obviating the need to wait for tests or hope that an antibiotic used before test results are available will be appropriate;
- in addition to destroying microorganisms through its specific action, BI boosts the overall immunological defenses of the body through its non-specific action, unlike chemotherapies; and
- BI therapy is less expensive than many chemotherapies.

While the handiness and precise dosages of antibiotics make them better than BI for some specific indications, in major indications such as pre- and postoperative prophylaxis, BI is superior.

How does BI compare with other therapies such as surgery and natural medicine?

BI is less expensive, damaging, and invasive than most surgery, while it can perform dramatically in acute conditions for which most kinds of natural medicine are not suited. In effect, BI shares certain characteristics with both mainline and alternative therapies. BI is like mainline therapies in that its mechanism of action is understood (at least by readers of this book!) and closely parallels that of drugs. In fact, its mechanism of action is understood much better than those of many competing treatments, and it helps elucidate the mechanisms of yet others. In addition, BI is in use among reputable physicians; it has been administered to millions of patients without evidence of significant, systematic negative side effects; it has been extensively reported in the medical literature; and the UBI device has FDA clearance.

BI is like alternative therapies in that in the United States only a few physicians use it. There is widespread ignorance and consequent unwarranted skepticism in regard to it. There were only a few old controlled clinical trials. And UBI has the allure of a natural therapy because UV irradiation is associated with sunlight.

So BI is a panacea of sorts?

No. In some disorders BI's effectiveness is known to be limited. In other cases, e.g., gastric ulcers, effective and relatively inexpensive treatments already exist, so there is little need even to investigate using BI--except as an option in a remote, impoverished region where modern medicines are hard to come by. In other indications, such as sinusitis, the results with BI appear to be good enough to make it worth trying in a given refractory case but not good enough to recommend as a standard treatment. BI is almost always ineffective against cancer.

One way to think about it is this: for what indications is BI the treatment of choice? From this perspective, BI is indeed unusual. No other therapy can match the range of difficult conditions for which BI is the clear or potential treatment of choice. Its mechanisms of action and therapeutic profile are well characterized. As a photobiological and immunological treatment, Blood Irradiation therapy is inherently attractive.

Chapter 4: Blood Irradiation Therapy for Infectious Diseases

Infectious diseases were the original target of UBI--this in an era before the advent of antibiotics when there was no other protection against infections. The first two treatments were of patients moribund with septicemia following septic abortions.

Knott and his medical colleagues soon came to understand that the known germicidal action of direct UV irradiation had little or nothing to do with the effects against infectious agents that UBI obtained *in vivo*, which were considerably greater than those possible through direct irradiation.

The secondary irradiation emitted by the irradiated cells concentrates in the DNA and other vital structures of microorganisms, causing strand breaks, pyrimidine dimers, and other disruptions that ultimately destroy them. In addition, BI boosts phagocytosis of microorganisms by white blood cells as well as humoral activity against infections; and it stimulates hematopoiesis in the marrow to produce a new set of immune cells. This multiple action helps explain why the effects can take place with dramatic swiftness--reversing serious conditions within a matter of hours--or they can emerge over days and weeks following BI.

From research on the germicidal effects of UV irradiation in the purification of water and other fluids, certain patterns of its action against microorganisms *in vitro* have been established:

> In general, most bacteria are killed by exposures to ultraviolet radiation ranging between about 16,000 and 20,000 microwatt seconds per square centimeter. In addition, viruses commonly found in drinking water supplies are killed by ultraviolet radiation

exposures ranging between about 6,000 and 40,000 microwatt seconds per square centimeter. Mold spores require up to 60,000 microwatt seconds per square centimeter in order to be killed, while parasites may require up to 200,000 microwatt seconds per square centimeter or more of ultraviolet radiation before such parasites are killed.[58]

In assessing the effects of BI on infectious diseases, several considerations must be kept in view. First, the action of BI depends largely on the concentration of secondary radiation--in both UV and visible wavelengths--emitted by the activated blood cells *in vivo*, and so it is superior to the effects of direct UV irradiation. Second, through mutations microorganisms could in theory reduce their vulnerability and possibly develop resistance to BI. There are no reports in the literature, however, that any microorganism has developed drug resistance to BI; this must be considered a priority question for further investigation. Third, the clear implications of these data regarding destruction by direct UV irradiation of microorganisms in water and other fluids outside the body are that bacteria and viruses are relatively more vulnerable than fungi and parasites, while considerable variation exists within each grouping.

Bacterial Infections

Streptococcal infections of many kinds have been treated successfully with BI. The only exception is subacute bacterial endocarditis with an associated streptococcus viridans bacteremia or septicemia.[59] In these cases, BI will successfully remove the infection but fail to undo the damage to the heart. Other streptococcal infections such as strep throat, rheumatic fever, scarlet fever, acute tonsillitis, acute otitis media, and erisypelas were reported by the pioneer American UBI practitioners to be very responsive to BI treatment.

In the area of staphylococcal infections, skin infections such as furuncles and carbuncles respond very well to BI. Acute osteomyelitis is

---

[58] John R. Noll and Stephen V. Montvila, U.S. Patent Number 4,769,131 (1988)
[59] Miley (1997), p. 54

another indication. In chronic osteomyelitis, the effect can be positive also even though less marked. A 1991 Russian study of 55 cases of chronic osteomyelitis used LBI as part of combination therapy including antibiotics. The researchers administered 1 to 3 sessions of LBI at 4 milliwatts for 20-25 minutes. They found that LBI helped to restore a balance among immune cells and considered that this contributed to good outcomes in most cases of subsequent surgery.[60] BI has also been used successfully in enteritis and colitis.

A Russian study of the use of UBI with 40 infants and children up to 3 years of age suffering from acute pneumonia and pleural infection found that they improved more rapidly than 25 historical controls.[61] Statistical problems with this study, however, make its claim that the use of UBI reduced mortality fourfold without foundation. Another study of 56 children up to age 1 with acute pneumonia measured against 45 controls treated with standard drugs discovered that the UBI-treated children's temperatures and rapid heartbeats dropped more swiftly; the morphological condition of their peripheral blood and the level of phagocytosis showed more improvement; and their hospital stays were reduced by 24 percent compared to the controls.[62]

In a study of 50 adults who contracted pneumonia in conjunction with skull and brain injuries, other researchers noted that 6-8 sessions of UBI plus endolymphatic antibiotics significantly raised both the number of T-cells and the levels of IgA and IgM over those of 25 controls who received standard antibiotic therapy.[63] The treatment group recuperated more swiftly than the controls. 8 (16 percent) of the first group died; 6 (24 percent) of controls did ($p<0.05$). Here, of course, the problem is that it is impossible to differentiate the effects of UBI from those of endolymphatic antibiotics.

A retrospective American study of the effects of UBI in the treatment of staphylococcus albus and staphylococcus aureus infections reviewed the results in 7 cases that ended in death and 9 that ended in

---

[60] Gostishchev, V.K. et al. "Effect of Intravascular LBI on the Immune System of Patients with Chronic Osteomyelitis [Russian]," Khirurgiia (1991), No. 9, pp. 98-101

[61] V.N. Kalinkin et al. "Autotransfusion of Blood Treated by Ultraviolet Irradiation in Destructive Pneumonia in Young Children [Russian]," Khirurgiia (1991), No. 8, pp. 14-20

[62] F.S. Shamsiev et al., "The Efficacy of UBI in the Combination Therapy of Acute Pneumonias in Very Young Children [Russian]," Pediatriia (1990), No. 11, p. 112

[63] A.B. Kibirev et al., "UBI and Endolymphatic Antibiotic Therapy in the Treatment of Pneumonia in Patients with Skull-Brain Trauma [Russian]," Zhurnal Voprosy Neirokhirurgii Imeni N.N. Burdenko (1990), No. 3, pp. 11-14

complete recovery. In 6 of the 7 fatal cases, intensive sulfa treatment had preceded or followed UBI treatment. In the seventh case, the underlying disease was bladder carcinoma. In 8 of the 9 patients who recovered, UBI was used alone; and in the ninth case only a small amount of sulfa drug was used. In effect, the combination sulfa and UBI group served as a control, leading the researcher to conclude that the administration of sulfa drugs with UBI was very dangerous, while UBI monotherapy was highly effective against staphylococcemia. He ascribed the deleterious effects of sulfa drugs to their toxic action; it is, however, more likely that they were photoactive drugs and that UBI activated them.[64]

E. coli and Other Infections

An American study of 7 cases of Escherichia coli septicemia found that UBI treatment saved 5 patients' lives. A sixth died of myocardial degeneration but with a sterile bloodstream at autopsy. In the seventh case, UBI treatment succeeded in fully suppressing the E. coli infection, but upon autopsy Staphylococcus aureus haemolyticus was found in the bloodstream. The researcher commented that in his experience S. aureus septicemia did not respond well to UBI.[65]

In a Russian clinical trial of patients with pneumonia, however, UBI was successful against pneumococcus, staphylococcus, streptococcus, and mixtures of microbes alike. 92 patients treated with standard drugs plus 5-7 sessions of UBI 1-2 days apart were compared with 90 controls (including 20 healthy individuals) who received standard drugs only. The physicians irradiated only 80-120 ml of blood each time, which is considerably below the Knott Technique's level but typical of Russian practice.

The treatment group recuperated more rapidly (by 5-7 days), had fewer complications, experienced a reduction in fibrogen to normal and an activation of anticoagulatory and fibinolytic elements by the end of the 2-3 sessions, and--in the cases of those patients with initial anemia--a

---

[64] George Miley, "Efficacy of Ultraviolet Blood Irradiation Therapy in the Control of Staphylococcemias," American Journal of Surgery 64 (1944), pp. 313-22

[65] E.W. Rebbeck, "Ultraviolet Irradiation of Blood in the Treatment of Escherichia Coli Septicemia," Archives of Physical Therapy 24 (1943), pp. 158-67 and 176

30.7 percent increase in erythrocytes by the time of the third or fourth session (those with normal levels of erythrocytes maintained them).[66]

BI has repeatedly been shown to be highly effective in the treatment of peritonitis arising from E. coli infections. Among the beneficial results is the suppression of paralytic or adynamic ileus that arises from BI's effects on the autonomic nervous system.

Gonococcal and bacillary infections are reported to respond quite well to BI. An early three-arm retrospective study of the use of UBI in typhoid fever determined that UBI alone was more effective than UBI plus antibiotics and much more effective than antibiotics alone. In a study of 9 patients over 10 years, 3 who received UBI and sulfonamides became afebrile in 31 days on average whereas 3 who received UBI alone became afebrile in 16 days. Two who received sulfonamides only took longer still to become afebrile while a third died.[67] This is one of many pieces of evidence that underline an important point: monotherapy with BI can be more effective than the use of BI in combination therapy.

In a clinical trial of the effects of LBI on microcirculatory disorders in pediatric meningococcal infections, researchers compared 18 children 3 months to 3 years old with meningococcemia who received LBI in addition to standard therapy with 20 who received only standard therapy. The 18 children received 5-8 sessions of 20-30 minutes at 632.8 nm and 5-8 milliwatts intensity.

After the first session the researchers noted a clinical improvement in microcirculation, including a pinkening of the skin and a disappearance of perioral, periorbital, and fingernail cyanosis. Symptoms of infective-toxic shock quickly disappeared. By the end of LBI therapy, conjunctive microcirculation had improved by many measures; the conjunctive index of the LBI group was 2.22 +/- 0.50 compared to 4.87 +/- 0.36 ($p<0.001$) for the controls. The overall condition of the children including microcirculation and hemodynamic status improved 2-3 days earlier than with standard therapy. The researchers did not report on destruction of infectious agents.[68]

BI's wide spectrum of action makes it very handy in pre- and postoperational prophylaxis against infections of all kinds. A Russian

---

[66] A.D. Novgorodtsev and E.M. Ivanov, "UBI as a Method of Nonspecific Therapy of Acute Pneumonia [Russian]," Voenno-Meditsinskii Zhurnal (1992), No. 12, pp. 38-39

[67] E.W. Rebbeck and H.T. Lewis, "Use of UBI in Typhoid Fever," Review of Gastroenterology 16 (1949), No. 8, pp. 640-9

[68] Brill' (1994), pp. 149-51

study of its effects in the treatment of patients following operations with bleeding in the central organs of the body is illustrative. The researchers compared 28 patients operated on for trauma in parenchymal organs, shock, and at least 1 liter of blood loss through internal bleeding. The patients received 3-7 sessions of LBI (632.8 nm) at 4-5 milliwatts starting immediately after the operations and continuing on a daily basis. 32 patients with equivalent injuries and operations served as a control group.

In the group that did not receive LBI, only 28.1 percent did not experience complications. In the LBI group, 67.9 percent had no complications.[69]

Tuberculosis

While neither the Americans nor the East Germans reported on experience in the use of BI against tuberculosis, recent Russian clinical trials have targeted tuberculosis of the lungs. In one, 222 hospitalized patients with destructive tuberculosis of the lungs were divided into a group of 86 who received standard antibiotics plus electrophoresis and 5 sessions of UBI, and a group of 136 who received antibiotics only. The antibiotics included izoniazid, rifampicin, streptomycin, and other drugs.

Within three months of the start of therapy, 100 percent of the first group were disease-free, while only 58.8 percent of the second group were. After six months, 86 percent of the second group had become disease-free; after eight months, 97 percent were disease-free. 3 percent remained infected.

After three months, destructive results of the disease disappeared in 89.5 percent of the first group but only 38.2 percent of the second ($p<0.001$). After six months, destructive results had disappeared in 90.7 percent of the first group and in 63.9 percent of the second ($p<0.001$). The average hospital stay of the first group was 48 days shorter. How the credit for these excellent results should be divided between electrophoresis and UBI is not clear.[70]

---

[69] Ibid., pp. 129-31

[70] V.Z. Zhandov et al., "Efficacy of Chemotherapy in Combination with Electrophoresis and UBI in New-Onset Cases of Destructive Pulmonary Tuberculosis [Russian]," Problemy Tuberkuleza (1995), No. 3, pp. 20-22

In a trial that included 119 tuberculosis patients, apparently without controls, LBI was used in combination with standard drugs in preparation for surgery. LBI was administered in 10-12 sessions of 10-15 minutes at 1-1.5 milliwatts. 105 (90.5 percent) of the patients were observed to have a stabilization of temperature, cessation or diminution of coughing, reduction in the amount of mucous produced, improvement of functional indicators of pulmonary ventilation, and a stabilization of T and B lymphocytes. The researchers concluded that LBI was a useful tool in preparing patients for operations, but their apparent failure to include controls made it hard to distinguish the effects of LBI from those of the drugs they were using.[71]

In a third trial, 88 tuberculosis patients were compared with an unspecified number of controls. 79.5 percent of the treatment group had accompanying illnesses of another organ, and 25 percent had complications of tuberculosis. The treatment consisted of 3 sessions of UBI (a low dosage). 5-6 days after the last one, the results were tabulated: 31.9 percent of those receiving UBI experienced significant improvement, 47.8 percent partial improvement, and 20.3 percent no improvement. Those who did not improve tended to have fibrous-cavernous tuberculosis, were repeat cases, had undergone operations on account of tuberculosis of the lungs, or were chronic alcoholics. The UBI treatment also had significant effects on such accompanying illnesses as chronic bronchitis, heart pain, diabetes, and high blood pressure.[72]

Viral Diseases

The pattern of effectiveness of Blood Irradiation therapy in the treatment of viral diseases parallels that in bacterial diseases. A major distinction is that antibiotics are competitive with BI in the treatment of bacterial diseases, whereas BI has shown itself to be effective against

---

[71] V.T. Dobkin and G.B. Bondarev, *New Applications of Laser Medicine* [Russian]. St. Petersburg, 1993, pp. 70-71, cited in E.V. Kul'chaveniia, "The Use of Low-Intensity Lasers in Phthisiatry [Russian]," Problemy Tuberkuleza (1995), No. 4, pp. 19-21

[72] R.G. Mingalimova et al., "UBI in the Complex Therapy of Patients with Tuberculosis of the Lungs [Russian]," Problemy Tuberkuleza (1995), No. 3, pp. 27-28

certain viruses against which antiviral drugs have only weak or no action.

The studies on which this conclusion is based include isolated case reports, long series of consecutive case studies, cumulative reports drawn from the cases of many practitioners, and clinical trials with full controls. "Anecdotal" evidence of the effectiveness of BI in treating a wide range of viral diseases that would otherwise be easy to dismiss gains in power by the way it dovetails with the harder evidence of clinical trials in a few such diseases. The fact that viral DNA is an easy target for the concentrated effects of the secondary radiation of BI adds to the perception that BI is a formidable treatment of viral diseases.

An early indication for ultraviolet BI was poliomyelitis, including polioencephalitis. Miley treated 58 cases in an epidemic of acute poliomyelitis in Los Angeles in 1943. While in the natural course of the disease the toxic symptoms could resolve themselves, the use of UBI might have also played a role in suppressing them. In 11 cases of bulbospinal poliomyelitis in which UBI was employed, 1 patient died, a mortality rate of 9 percent compared with the 35 percent rate in similar cases in this epidemic where UBI was not used. Identical results were obtained in a more closely controlled study at Children's Hospital in Washington, D.C.: of 25 cases of bulbospinal poliomyelitis, 11 were treated with UBI, of which only 1 died. Among the 14 not treated with UBI, 5 died (35.7 percent).[73]

Many cases of encephalitis have been successfully treated with BI, though severe cases sometimes do too much damage for BI to be of assistance. The same pattern holds for the treatment of influenza and common upper respiratory tract infections. BI is likewise reported to be very effective in the treatment of herpes zoster. Reports on the treatment of herpes simplex are mixed. The results of BI on mumps and measles are said to be uniformly excellent. BI is also reportedly effective in treating mononucleosis (EBV) and plantar warts (*verucca plantaris*--a human papilloma virus akin to those that cause genital warts and that have been implicated in the genesis of cervical cancer).

In a report entitled "Ultraviolet Blood Irradiation Therapy in Acute Virus and Virus-Like Infections,"[74] the researchers listed 79 consecutive cases of various virus infections--viral pneumonia, poliomyelitis (spinal and bulbospinal), herpes simplex, herpes zoster,

---

[73] Miley (1997), pp. 31-33

[74] George P. Miley and Jens Christensen, Review of Gastroenterology 15 (1948), No. 4, pp. 271-83

herpes zoster ophthalmicus, and fulminating mumps--treated with UBI. These ranged from early through moderately advanced to "apparently moribund." 78 of the patients recovered, including 8 of 9 apparently moribund with viral pneumonia or bulbospinal poliomyelitis.

The researchers concluded:

> Inasmuch as there exists no other method of favorably influencing or controlling virus or virus-like infections, ultraviolet blood irradiation therapy, in the light of experience to date, deserves a wide application in the treatment of such infections.[75]

G.J.P. Barger, M.D. of Georgetown University Hospital commented on the report that he had administered over 2,500 treatments of UBI, including 15 for viral pneumonia and 29 for poliomyelitis, and that he concurred with the findings. He had successfully treated three pediatric viral encephalitis cases as well--one of meningeal encephalitis and two of encephalitis lethargica, he added.

Ukrainian Findings

In 1996, physicians in Odessa reported on a trial of intravenous laser BI in the treatment of 200 women infected with herpes simplex viruses 1 and 2.[76] 120 of the patients were pregnant. There were no concurrent controls, though an historical control was apparently used. The patients received an unspecified number of treatments with the standard helium-neon red laser at 632.8 nm, 4 milliwatt strength, for 20-25 minutes. LBI was administered in addition to the ordinary drug treatment.

The researchers found that the efficacy of treatment was increased by 35 percent (measure unstated) over drug therapy alone. It was possible to reduce the dosage of drugs to 1/3 normal (in other words, the standard drug treatment was used, but at a much lower dose than is ordinary), and remissions lasted 10-12 months (no comparisons or

---

[75] Ibid., p. 276

[76] Ministry of Health of Ukraine (1996), pp. 178-81

explanations were provided). None of the babies subsequently born to the pregnant women showed any pathology that might have been caused by herpes infection.

The inadequacies of the design and reporting of this trial detract from its formal validity, but the data provided are nonetheless suggestive.

In another trial of 33 patients with genital infections (12--staphylococcus, 7--epidermal staphylococcus, 2--enterococcus, 12--herpes simplex), LBI at 670 nm was given at the very low dosage of 0.1 milliwatts for an average of 10 sessions. 10-15 days after the end of therapy, no evidence of bacteria was found in the blood of 15 out of the 19 patients with bacterial infection. Of the 12 herpes patients, no evidence of virus was found in the blood in 7 patients. Again, the lack of controls and of detailed reporting make this study interesting but lacking in definitiveness.[77]

A Russian study found UBI to be very effective in two cases of severe Highmoreitis with complications in children following viral infection of the facial sinuses.[78]

Viral Pneumonia

The early American UBI practitioners considered UBI highly effective against all types of pneumonia, regardless of origin. They noted that it rapidly resolved cyanosis so that patients could be taken off oxygen, then the clearing up of the lungs took place over a matter of days--more slowly if an antibiotic was used in conjunction with UBI. In particular, they reported that a single treatment with UBI was often sufficient to bring about recovery from viral pneumonia.[79]

In a Russian study of 10 male patients, aged 17-30, with acute pneumonia and 20 healthy controls, all subjects were given standard antibiotics and other drugs as well as a two treatments with UBI. The researchers found that UBI had a markedly favorable influence on the

---

[77] Ibid., pp. 176-8

[78] Yu. Ya. Ozerskii et al., "UBI in Severe Cases of Highmoreitis [Russian]," Vestnik Otorinolaringologii (1995), No. 1, pp. 49-50

[79] Miley (1997), pp. 75-78

regularization of lipid peroxidation and the antioxidant system, with overall beneficial results for uneventful recovery.[80]

A Russian study of the comparative effects of LBI and standard drugs on acute pneumonia found that fever and intoxication disappeared an average of 4.8 days sooner in the LBI group and that auscultatory symptoms disappeared 3.29 days sooner. The result was that the LBI group had much less ongoing damage to the lungs. The patients had virtually no perivascular suppuration or erythrocyte aggregation as well as little of the disruption of the connection between arterial and venous blood characteristic of some 70 percent of the standard drug group.[81]

Hepatitis

In regard to the treatment of hepatitis with BI, several considerations must be kept in mind. First, the term "hepatitis" is applied to a wide range of acute and chronic liver disorders of varying origin and natural course. Second, during the period of UBI's widest use in the United States (up to 1955), many aspects of hepatitis were poorly understood. Third, most cases of acute hepatitis are self-healing. Fourth, antibiotics have never been particularly effective in the treatment of hepatitis. Fifth, the Russians have reported almost no clinical findings regarding the treatment of hepatitis with BI. However, in a study of 64 patients and 20 controls, transdermal BI (890 nm) was reported to be successful in terminating acute hepatitis 4-5 days earlier than in controls. Its action against the few cases of chronic hepatitis B and C treated was less clear.[82]

In the US there was one major article by R.C. Olney and 15 colleagues that appeared in the *American Journal of Surgery* in 1955.[83]

This report included 43 cases to which an average of 3.28 UBI treatments were administered. In all of the 43, the acute attack quickly

---

[80] V.A. Yaklovlev, V.V. Vyitrishchak, and M.A. Kharitonov, "UBI in Acute Pneumonia: Some Mechanisms of Therapeutic Action and Optimization of Intervals between the Procedures [Russian]," Terapevticheskii Arkhiv (1994) 66, No. 8, pp. 39-42

[81] Use of Low-Intensity Lasers in Experimental and Clinical Medicine [Russian]. Izhevsk, 1994, pp. 63-64 (hereafter Izhevsk)

[82] Ovsiannikov (1997), pp. 40-41

[83] R.C. Olney et al. (1955), pp. 402-9

resolved; but this could well have been owing to the natural course of the disease. Still, the use of UBI may have kept certain of the acute cases from evolving into chronic ones. More important, of the three clearly definable chronic cases in the series (#39, 40, and 43), all responded well to UBI, though one required more UBI treatment for a recurrence. None of the three thereafter had any complaints during follow-up of 6, 4 1/2, and 6 years, respectively. In other words, UBI appears to have been very successful in treating chronic hepatitis; and this fact was observed and reported by respectable physicians in a standard medical journal.

A Russian trial investigated the effects of using LBI in the portal vein in the treatment of diffuse liver disorders (chronic active hepatitis and cirrhosis of the liver) following liver surgery. Of 46 patients, Group 1 (20 patients) received drug infusion into the portal vein; Group 2 (16 patients) received standard intravenous LBI; and Group 3 (10 patients) received infusion therapy with intraportal LBI. In Group 1, 12 of 20 had good results, while 2 died of liver insufficiency. In Group 2, 13 of 16 had good results. In Group 3, 10 of 10 had good results. The researchers thought that improvement of microcirculation in the liver accounted for the superior outcomes of Group 3.[84]

Eye Infections

BI has important indications in ophthalmology, including in the treatment of viral and T-cell-driven inflammatory disorders. It was used by the early American UBI practitioners to treat a range of eye infections. In a study that included a group of 10 controls, ophthalmologists treated 27 patients with 3-7 sessions of UBI.[85] Many of them had been refractory to drug treatments. The control group was treated with antibiotics, Cortisone, Antigen-H, and triple typhoid vaccine.

The researchers commented:

---

[84] Izhevsk (1994), pp. 109-111

[85] Donald F. Farmer, Clifford P. Sullivan, and William B. Sullivan, "The Use of Hemo-Irradiation (Knott Technique) in Eye Infections," Industrial Medicine and Surgery 21 (1952), pp. 183-85

All cases treated with UBI responded exceptionally well, except for one, an irido-cyclitis, which also failed to respond to conventional therapy....the cases treated with UBI were singularly free of sequelae, such as posterior synechiae (adhesions of the iris to the lens), exudates on the anterior lens capsule, heavy keratic precipitates, with the attendant loss of vision. The above-mentioned sequelae were present, and of a permanent nature, in 30 % of the control group.

The authors provided a table of their cases:

Table 5.1. UBI Treatment of Eye Disorders

Type of Infection	Total cases treated	No. treated with UBI	No. treated conventionally	No. treated with combined therapy
Irido-cyclitis	28	18	10	
Uveitis complicated by glaucoma	3	3		
Retro-bulbar neuritis	3	3		
Herpes zoster ophthalmicus	1	1		
Keratitis	1	1		
Sympathetic ophthalmia				1

One severe case of sympathetic ophthalmia (a kind of uveitis) was treated with Cortisone and UBI; all others in the treatment group received only UBI. The UBI cases were discharged in an average of 17.5 days vs. 30.8 days for the control group.

HIV

The excellent track record of BI in the treatment of other viruses would appear to make it a promising candidate to be an inexpensive, low-toxicity, high-compliance, and effective treatment of HIV.

One finding that misled HIV researchers was the discovery that ultraviolet irradiation activates HIV. In fact, in one of the most detailed studies of this phenomenon, the authors themselves noted that at a higher dose the destruction of virus would occur.[86] In effect, their study regarding HIV activation was a special case that suggests what happens inside the body when the secondary radiation from the irradiated blood cells first encounters the HIV virus. It rapidly activates viral DNA while at the same time beginning to disrupt it (with formation of pyrimidine dimers, etc.). Then, as the dose of secondary radiation increases, the concentrated secondary radiation proceeds to destroy the DNA. In addition, the studies that identified the activation effect that UV radiation had on HIV were *in vitro* ones, while as early as the 1930s Knott had realized that the basic mechanism of UBI was an *in vivo* phenomenon.

Various American UBI practitioners have treated HIV patients. However, in 1997 the FDA seized one company's UBI devices that were being used to treat HIV on the grounds that this was not an indication at the time the device was accorded legal status in 1976. There is no evidence that the FDA considered the question of whether, regardless of the legal merits of the case, UBI was in fact highly effective.

A few Russian physicians have used LBI on HIV, and there are several intriguing reports.[87] Anecdotal reports on the effectiveness of BI are mixed, and only a few have made their way into print.[88] One German source claims the disappearance of HIV from the blood following Hematogenic Oxidation Therapy (HOT: UBI plus blood oxygenation) of a 24-year old woman.[89] The pattern that emerges is reminiscent of BI's record in certain other diseases. Many of the HIV patients who turn to UBI practitioners have advanced or even terminal disease, with hundreds of billions of copies of virus and devastated

---

[86] B. Zmudzka and J. Beer, "Activation of Human Immunodeficiency Virus by Ultraviolet Radiation," Photochemistry and Photobiology 52 (1990), No. 6, pp. 1153-62; see p. 1161.

[87] For instance, a report of improvement in 6 HIV patients with moderately advanced disease treated with a low-intensity laser aimed at stimulating the marrow through the skin: Victor Ovsiannikov and N. Sizova, "The Treatment of AIDS-Patients by Means of Laser Action," International Conference on AIDS, July 7-12, 1996; 11(2):88 (abstract no. We.B.3182). Ovsiannikov (1997) reports that transdermal BI at 890 nm led to significant rises in CD-4 counts in two patients over a 9-month period (369 to 520 and 103 to 467). 8 out of 8 patients had good responses to therapy.

[88] Miley (1997), pp. 233-240; Douglass (1994), *passim*

[89] Hans Stadtlaender and Klaus Lippmann, *Woerterbuch der HOT*. Heidelberg: Karl F. Haug Verlag, 1988, p. 124

immune systems; and so UBI has little effect on their condition, and that is palliative at best. There are, however, oral reports that UBI can clear all evidence of virus from the blood. If accurate, these must be presumed to refer to early-stage HIV.

In Russia some physicians consider HIV a counterindication for BI because the damage BI can cause to the membranes of T-4 cells make them more vulnerable to subsequent infection by HIV. How this would occur if BI treatment successfully removed all but a few scattered remnants of HIV from the body is not clear, and their judgment is open to doubt.

With HIV as with many other conditions, the correct question would appear to be not "Can BI treat this disorder?" but rather "What is the stage of this disorder?" If it is a relatively early stage, then it is likely that BI can be effective because BI relies on the underlying order of the body to achieve its pharmacological effect. Once that order is undermined by the progression of a disease to an advanced stage, it becomes much less likely that BI can be effective--though there are reports of it curing diseases of many years' standing. These cases appear to be mainly chronic inflammatory disorders affecting a single organ and so not necessarily distorting the order of the entire organism.

The history of the treatment of HIV with various chemotherapies suggests that this principle is also operative with them. The recent shift to the use of the 3-drug combination including protease inhibitors in early-stage HIV is a correct response to the realization that "waiting and watching" is a disastrous policy in HIV as in many other diseases.

Thus it must be anticipated that BI is potentially very effective in the treatment of early-stage HIV. BI's track record in the treatment of other infectious diseases suggests that it would be as effective as the protease inhibitors in reducing virus in the blood to below detectable levels. Less clear, however, is whether the virus could develop resistance to BI. No reports of such resistance to BI among other microorganisms exist. Still, HIV has a high capacity for mutation, and so the only way to answer this question will be to test BI in clinical trials against HIV.

It is to be expected that BI will be more effective than protease inhibitors in the treatment of many HIV-related conditions. In particular, it should be very effective in the treatment of hepatitis, pneumonia, and various opportunistic infections. It is said to be very effective in treating CMV, for instance. It would be useful in the treatment of acyclovir-resistant herpes infections. BI may also be superior to the protease inhibitors in suppressing the perinatal transmission of HIV and possibly in preventing the infection of fetuses or nursing infants with maternal HIV. It also appears to be a prime candidate for the treatment of such sequelae as ITP. It is unlikely, however, that BI can have a significant effect in Kaposi's sarcoma.

The relatively low cost of BI would make it ideal for the treatment of poor HIV patients in developing countries--the great majority of new HIV cases. BI also has the advantage in terms of compliance that it can be administered in a few sessions in a clinic, and any failure to appear will at least be known to the medical practitioner--unlike the compliance problems that result when patients do not take their medications. The simplicity of the BI administration regime compared to the complicated dosing of the protease inhibitors is another advantage.

From the perspective of BI, HIV might be just one more virus.

Fungi and Protozoa

In theory, BI should prove at least somewhat effective in the treatment of both fungal and protozoan/parasitic infections. It has, for instance, been used against malaria. According to one report, good results were obtained in the treatment of 6 cases out of 7 of innoculation malaria; but there are no thorough studies of BI's effects on any of these other infections.[90] A Russian trial of the use of UBI in candidiasis resulted in the cure of 8 out of 10 cases.[91]

Deactivation of Toxins

A curious, not-yet-explained property of BI is its ability to deactivate toxins. That sunlight and UV irradiation can deactivate snake venom and other toxins *in vitro* has long been known.[92] BI appears to have the same action, e.g., in the deactivation of botulinum. A single

---

[90] Miley (1997), pp. 225-6
[91] Frick (1989), p. 75
[92] Douglass, pp. 41-48

treatment of UBI was reported to have sufficed to bring a woman from an advanced botulism coma back to full health within 13 days.[93] One way in which BI might deactivate toxins is by disintegrating clumps of fat particles--cholemicrons--in the blood so that their normal Brownian movement is restored. Cholemicrons are known to assist in deactivating toxins.

It is likely that certain of the beneficial effects of BI in the treatment of staphylococcus, typhoid, and other infectious agents arise from its ability to deactivate toxins.

BI is likewise able to deactivate the chemical warfare agent ricin. Further study of this deactivation effect might yield insights regarding the action of BI in other conditions.

Conclusion

In sum, BI appears to be safe and effective in treating dozens of infectious diseases. There is no evidence that microorganisms can develop resistance to it, though no thorough study of this question has been undertaken. In some circumstances--ulcers, for instance--antibiotics are clearly simpler and even cheaper than BI therapy. In many others, however, BI can be competitive with antibiotics.

BI possesses major advantages over antibiotics. It can successfully treat certain viral diseases that are not responsive to antiviral drugs. It is enormously versatile and therefore appropriate for clinics in developing countries that cannot afford an arsenal of antibiotics. BI's versatility is also highly useful in cases when early intervention, even before a precise diagnosis of the infectious agent can be made, is of the essence. BI has also frequently saved lives in cases of septicemia when all drug therapies have failed. In view of the high cost of some antibiotics and the growing problem of multidrug resistance, a thorough investigation is warranted of the potential of BI for the treatment of infectious diseases of all kinds.

---

[93] George P. Miley, "Recovery from Botulism Coma Following Ultraviolet Blood Irradiation (Knott Technic)," Review of Gastroenterology 13 (1946), No. 1, pp. 17-19

Chapter 5: UBI and LBI in the Treatment of Bronchial Asthma

Bronchial asthma is a widespread, debilitating ailment of modern civilization. In the United States, some 5-7 percent of adults and 7-10 percent of children are estimated to have BA. It is the leading cause of chronic illness in children. BA arises from various causative factors--allergies to household dust (mites), to animal dander, to molds, to pollen, to smoke, and to certain foods and food additives as well as industrial and automotive pollution and workplace pollutants. Idiopathic BA can follow from common upper respiratory viral infections, presumably as the immune system attacks cells that have been altered by virus so as to appear foreign to the body, which explains why many successful initial treatments need to be followed by regular maintenance in order to deal with new accumulations of new immune cells. BA can also have a strong psychological component, with attacks triggered by emotional upsets.

BA causes wheezing, coughing, dyspnea, sleeplessness, and much general misery; and the toxic effects of various drugs used to control it can add to the problem. While there is some dispute about BA's precise mechanism of action, it is clear that hypersensitive immune cells play a major role in the exacerbations. The results are narrowing of air passages from bronchospasm, mucosal edema, and mucosal secretions. In turn, the deterioration in air flow can lead to a severe, unrelenting attack known as status asthmaticus, which can be fatal. In the United States, some 750 fatalities per year are ascribed to asthma.

Scores of therapies have been tried against BA over many decades--some more successful than others but none with truly satisfactory results. At present a battery of drugs is employed with considerable success in relieving individual attacks, but with toxicity, expense, and often disappointing effects on the course of the disease. Improved understanding of the autoimmune basis of many cases of BA has led to the use of methotrexate, which offers the hope of better outcomes; but methotrexate has its own toxicities and must be used on a continuing basis.

So it is not surprising that BI practitioners have employed their therapy against BA. The results are important in many ways: first, they demonstrate a noteworthy application of BI; second, they highlight the methodological idiosyncrasies of the early studies by American proponents of UBI; third, they show the strengths and weaknesses of the methodology of the Russian BI researchers; fourth, the classic experiments conducted by the Russians on BA led to the first strong evidence regarding the relative merits of UBI and LBI; and fifth, this evidence gives a clue to an fascinating mechanism whereby BI attains its results.

The American Studies

George P. Miley, M.D. and his associates reported in 1943 on a series of 80 consecutive cases of intractable BA treated between 1938 and 1942.[94] The researchers defined "intractable" according to these criteria:

- failure to respond to removal of known aggravating extrinsic factors, e.g. dusts, pollens and foods.
- Refractoriness to all desensitization efforts.
- Refractoriness to bronchoscopy and bronchoscopic "drainage" with or without supplementary use of autogenous vaccine.
- Failure of intranasal operations performed for the relief of nasal obstruction or as a treatment of chronic sinus infection to relieve asthmatic symptoms.

---

[94] G.P. Miley, R.E. Seidel, and J.A. Christensen, "Preliminary Report of Results Observed in Eighty Cases of Intractable Bronchial Asthma," Archives of Physical Therapy 24 (1943), pp. 533-42

- Temporary or no relief following inhalation or injection of epinephrine or ephedrine preparations.

They administered UBI every 4-6 weeks until amelioration of symptoms over a 2-month period. Then they introduced maintenance therapy at 8-10 week intervals with the aim eventually of reducing the frequency to 3-4 treatments yearly. They also reduced epinephrine or ephedrine to a minimum.

There were no controls, a serious weakness in the study. Of the 80, 24 dropped out before they could be fully treated or followed up, which greatly detracts from the conclusiveness of the study. Still, 9 of the 56 treated were relatively symptom-free for more than one year; 20 were relatively symptom-free for more than 6 months; and 16 were relatively symptom-free for less than 6 months. 11 registered no response.

There was no clear explanation for the wide range of response. Patient # 3, a 30-year old female who had suffered from BA since the age of 3, with 25-40 severe attacks monthly at the time of the initiation of UBI treatment, underwent 4 UBI treatments over a period of 41 months. The end results were 1-2 mild attacks annually. In contrast, patient # 15, a 34-year old male with an 8-year history of BA and 30-50 attacks monthly with constant insomnia, showed no improvement whatsoever over a 12 1/2 month period in which he received 3 UBI treatments.

The researchers noted that during the first night after the initial UBI treatment a moderately severe attack often occurred, especially in the more advanced cases. Following subsequent UBI treatments, milder attacks occurred infrequently, diminishing over time.

Long-standing cyanosis slowly disappeared as the therapy progressed.

No side effects were observed.

The researchers were aware that the psychological component often observed in BA could account for positive responses to a new therapy, so they tested one case and reported thus:

> If a patient has had intractable bronchial asthma for thirty-eight years plus insomnia for six years despite the best of medical care, and has little or no asthma, plus freedom from insomnia, for

> almost four years from the time she first is given ultraviolet blood irradiation therapy, she obviously has had much more than psychotherapy. If the same patient has had, unknown to her, a withdrawal of the therapy (by closing of the rotating shutter of the Knott hemoirradiator and cutting off the source of ultraviolet rays from the returning blood) and begins to show signs of severe asthma again, she is suffering from a true ultraviolet deprivation. If we next really give the patient ultraviolet blood irradiation therapy and her symptoms once more clear up, we feel that this is proof positive that such a patient--and there is such a patient (patient 7)--not only has been benefited by the ultraviolet irradiation of blood but has been suffering for years from severe ultraviolet deprivation, which apparently plays a much greater part in this disease, as well as others, than has been suspected.

In the list of patients, #7 is entered as a 56-year old female with duration of BA of 17 (not 38) years. She was given 17 treatments of UBI and observed for 46 months. Her condition changed from "Severe asthma, insomnia for past 6 years, cyanosis" to "1-2 mild attacks monthly, no cyanosis, no insomnia, one seasonal aggravation."

Three comments appear in order regarding this interesting case:

- The blinded, double-reverse maneuver transformed this patient temporarily into a single control both for her own case and for the entire study. By demonstrating that reinstituting UBI therapy caused an improvement, the researchers were providing highly suggestive even if not conclusive evidence that the effectiveness of the treatment did not depend on a psychological effect, and that it was the UBI treatment and not any other cause that was improving her condition.
- The high number of treatments with UBI--17--suggests that other patients might also have improved more if they had received higher dosages of UBI over time. The highest number of UBI treatments any of the patients who failed to show any improvement received was 8.
- The "ultraviolet deprivation" theory was Miley's effort to make sense of the excellent results he was obtaining with UBI in many diseases. It has no scientific basis, however. The specificity with which BI targets the metabolically active immune cells to suppress or destroy them is a far better explanation of its mechanism of

action in BA. In addition, BI affects the autonomic nervous system, thereby relaxing muscles and opening up bronchial air passages. Its regularization of the various components if the blood also may play a role.

In a second study of 160 consecutive cases reported by the same researchers in 1946[95], 40 patients dropped out before complete follow-up. Using the same criteria and procedures as in the first study, the researchers discovered two important trends. The first was that the longer a patient remained under treatment, the better the results were. Of the 43 patients treated for less than 6 months, 15 (34.8 percent) responded favorably (remaining relatively symptom-free or improving definitely). Of the 44 treated for 6 months to one year, 32 (72.7 percent) responded favorably. Of the 17 patients treated for one to one and one-half years, 15 (88.2 percent) responded favorably. The other two improved slightly. Of the 16 patients treated for 2 to 5 years, 15 (93.7 percent) responded favorably. The other one improved slightly.

Of course, one explanation would be that those who did not benefit from UBI dropped by the wayside. Still, the statistical evidence suggests that there was genuine benefit for most sufferers from BA in persisting for years with UBI.

The second finding was that the younger the patient, the better the response to UBI was. Of 25 patients between 3 and 18 years of age, 16 became entirely symptom-free and 7 definitely improved, for a total of 23 (92 percent). Of 19 patients between 18 and 30, 11 became relatively symptom-free and 5 definitely improved, for a total of 16 (84.2 percent). Of 26 patients between 30 and 40, 7 became relatively symptom-free and 8 improved definitely, for a total of 15 (58 percent). Of 37 patients between 40 and 60, 5 became relatively symptom-free, and 15 improved definitely, for a total of 20 (53.9 percent). Of 13 patients over 60, none became relatively symptom-free and 3 improved definitely, for a total of 3 (23 percent).

These results suggest that UBI is especially effective against childhood BA.

In both studies the researchers sought to divide the results according to the types of BA, but there were no clear trends. Changes in

---

[95] G.P. Miley, R.E. Seidel, and J.A. Christensen, "Ultraviolet Blood Irradiation Therapy of Apparently Intractable Bronchial Asthma," Archives of Physical Medicine 27 (1946), pp. 24-29

our understanding of the etiologies and diagnoses of BA in the intervening 50 years reduce the value of this approach. The researchers did note that, as in other applications of UBI, those patients with less advanced disease tended to respond better than the advanced cases.

The early American UBI practitioners also used UBI against hayfever. Miley reported that in treating over 50 consecutive patients, he was able permanently to relieve all symptoms in about 1/3 of the cases and relieve symptoms with annual treatments in about 1/3 of the others. 1/3 experienced only mild or no relief.[96]

The Russian Studies

One of the main Russian centers of BI research is Vladivostok. In 1987 a book on UBI[97] was published there, and another on LBI[98] was published in 1993. In this second volume, the researchers report on clinical trials of BI in the treatment of hundreds of patients with bronchial asthma. BA is, of course, a serious problem in the former Soviet Union on account of high levels of industrial and vehicular pollution as well as cigarette smoking.

In the first study, 267 BA patients with BA related to infections and with moderately severe or severe course, including 93 with hormone dependency, were divided into two groups. Group 1 of 150 patients received fasting treatment only. Group 2 of 117 patients fasted and received UBI. The average fast was 14 days. Group 2 received 3 UBI treatments in the first two days of fasting.

Group 1 patients got over dyspnea on the seventh day on average and wheezing (by auscultation) on the twelfth. Group 2 got over dyspnea on the fifth day and wheezing on the tenth. 30 percent of Group 1 still had some clinical BA problems at the end of the treatment and required antibiotics. None of Group 2 had clinical problems. 25 percent of Group 1 had infections (frequent as a response to fasting), while only 11 percent of Group 2 had them. Some of Group 1 required antibiotics;

---

[96] Miley (1997), pp. 137-8

[97] S.S. Bolkhovitinova, ed. *Ultraviolet Blood Irradiation in Medicine* [Russian]. Vladivostok, 1987

[98] Sukhanova (1993); the following expands on pp. 34-36 above.

none of Group 2 did. The immune indicators for Group 2 were considerably better.

In a second study, the researchers made clear that they would be happy to find a justification for using LBI instead of UBI. The shortage of disposable transfusion sets in Russia, the expense and difficulty of obtaining heparin, and technical difficulties in carrying out the procedure (clotting in needles and in the cuvette, bad veins, etc.) were arguments against UBI.

So they ran a trial of 364 BA patients with moderate and severe disease. Group 1 consisted of 125 patients who received only fasting treatment. Group 2 had 122 patients who received fasting treatment and LBI. Group 3 included 117 patients who received fasting treatment and UBI.

The patients in Group 3 were given 3 UBI treatments over 2 days at the outset of fasting, with the amount of blood irradiated being quite low: only 1 ml per kg, or less than half of the level of the Knott Technique. The Group 2 patients received 3 LBI treatments over 2 days at the outset of fasting, at the level of 4-6 milliwatts for 15-20 minutes. The laser was a helium-neon one, presumably operating at 632.8 nm.

In Group 1, it took on average 8 days of fasting for incidents of breathlessness to disappear, whereas in Groups 2 and 3 this occurred at 5-6 days. In Group 1, it took 13 days for dry coughing to disappear, whereas in Groups 2 and 3 this occurred at 10 days.

Group 2 (fasting plus LBI) scored higher (37.5 percent) at the 3-day mark than Group 3 (fasting plus UBI) (27.7 percent) in the disappearance of incidents of breathlessness. The researchers attributed this to LBI's more rapid resolution of bronchospasm as a result of LBI's reduction in the level of histamine and serotonin in the blood. But after 7 days incidents of breathlessness had disappeared in 100 percent of Group 3 whereas they continued in 12.5 percent of Group 1 and 6.9 percent of Group 2. In other words, UBI had a more consistently powerful action than LBI.

At the time of release from the hospital, dry coughing had ceased in Group 3 whereas it continued in 13 percent of Group 1 and 11 percent of Group 2. Group 3 had superior dynamics of the indicators of external breathing (of an unspecified amount) and a lower quantity of infectious complications: Group 1 had 25 percent; Group 2 had 14 percent; and Group 3 had 11 percent.

Dividing the same set of patients into three different categories according to the type of BA led to more nuanced results. In allergic BA and infection-dependent (idiopathic) BA without the involvement of chronic bronchitis, fasting plus LBI worked more rapidly than fasting plus UBI, though it is not clear whether it was ultimately more effective. In infection-dependent BA with the involvement of chronic bronchitis, disappearance of obstruction and the inflammatory process in the bronchi occurred in 98 percent of the fasting plus UBI patients but only in 82 percent of the fasting plus LBI patients.

The researchers concluded that LBI's bronchodilating and hyposensitizing effects were superior to those of UBI whereas UBI's antimicrobial and anti-inflammatory effects were superior to those of LBI. Thus for allergic and idiopathic BA, LBI was to be preferred, whereas for BA with chronic bronchitis, UBI was preferable. They also argued that both types with fasting treatment were superior to drug treatments, but the lack of a control group receiving drug treatment robbed this conclusion of much of its force no matter how justified it might have seemed. In addition, a more sustainable conclusion regarding LBI might have been that it was to be preferred to UBI in cases where more rapid results were desired but that UBI provided more consistently excellent results.

A third trial examined the comparative impacts of UBI and LBI on the immune systems of 123 BA patients. While the researchers judged the results inconclusive, they did show that UBI tended to boost the absolute and relative (to T-8 cells) numbers of T-4 (helper) cells and reduce phagocytosis, while LBI tended to lower the numbers of T-4 cells and boost phagocytosis.

A fourth trial focussed on the comparative impacts of UBI and LBI on microcirculation in patients with severe infection-dependent BA in conjunction with chronic bronchitis. Of 202 patients, Group 1 consisted of 51 who received only fasting treatment; Group 2 had 80 patients who received fasting treatment plus LBI; and Group 3 had 71 patients who received fasting treatment plus UBI. All treatments were identical to those of the other three trials.

Using a sophisticated scoring system based on a combination of perivascular, venous, and intravenous factors, the researchers found that one week after the commencement of therapy, Group 2 (fasting plus LBI) made much more rapid improvement in both general and rheographic indexes.

In the second week, however, Group 3 (fasting plus UBI) forged ahead of the other groups. By the end of the third week, Group 3's indicators were normal whereas those of Group 1 and Group 2 lagged.

The researchers attributed the superior performance of UBI to its greater antiinflammatory action and the antimicrobial action of UV irradiation. They concluded that fasting with either LBI or UBI delivered a "beautiful" result, but that UBI was to be preferred in cases of infection-dependent BA in conjunction with chronic bronchitis while LBI was satisfactory for simpler cases.[99]

A Ukrainian study of LBI on 110 BA patients and 42 controls at a Yalta spa found that it conveyed measurable benefits to many patients when added to a full combination of natural and drug interventions compared with the control group which received the full combination without LBI.[100]

---

[99] The Russians have also used UBI and LBI in clinical trials on chronic bronchitis, chronic lung infections, cor pulmonale, and other respiratory syndromes. In one trial, 71 patients with primary and secondary bronchiectasis or abscessing pneumonia were divided into Group 1, 20 controls receiving standard drugs and fibrobronchoscopic lavage, Group 2 of 27 who received 3 treatments with LBI, and Group 3 of 24 who received 3-5 treatments with UBI. Group 3 had the most serious cases, including 6 with empyema. The outcomes of Group 2 and Group 3 were found to be equivalent and superior to those of Group 1. (Sukhanova (1993), pp. 46-51) In a 6-arm trial of cor pulmonale, fasting therapy plus either UBI or LBI were found to be superior to standard therapy plus UBI and LBI. (Ibid., pp. 69-71) In a third trial, 10 patients with chronic bronchitis and 11 with bronchial asthma were given 5-9 treatments of UBI. There were no controls. 16 obtained a marked all-around improvement. The researchers made note of the significant improvement in the structure and functioning of red blood cells, which they attributed to UBI's enhancement of erythropoiesis in the marrow. Many patients were able to go off prednisolone. Many also were able to shorten their stays in the hospital. (N.R. Paleev et al. "The Effectiveness of UBI in the Treatment of Chronic Nonspecific Pulmonary Disorders [Russian]," Vestnik Rossiiskoi Akademii Meditsinskikh Nauk (1993), No. 3, pp. 3-6). In a trial of LBI as a preparatory treatment for lung surgery in patients with chronic lung disease, 53 patients with pulmonary abscess, bronchiectasis, or defective development of lungs and bronchial tree were given 5 treatments of LBI for 45 minutes each. The control group of 51 received no LBI. Postoperative complications in the LBI group were half those in the control group, and the hospital stays were 8.1 days shorter (I.K. Marchuk and V.N. Kuzmich, "The Use of LBI in the Combination Treatment of Chronic Suppurative Pulmonary Disease [Russian]," Klinicheskaia Khirurgiia (1993), No. 6, pp. 9-10). Lastly, in treatment of 49 lung disease patients with UBI (9--acute pneumonia; 14--pulmonary abscess; 19--chronic bronchitis; 7--BA), good results were attained with all patients; but especially in the cases of chronic bronchitis and BA there was an initial exacerbation of symptoms before a healing effect took hold (Ganelina and Samoilova (1986), pp. 103-9).

[100] M.M. Yusupalieyeva et al., "LBI in the Spa Treatment of Patients with Bronchial Asthma [Russian]," Voprosy Kurortologii, Fizioterapii i Lechebnoi Fizicheskoi Kultury (1996), No. 5, pp. 9-12

Assessment

In assessing the admirable, groundbreaking work of the Vladivostok researchers, several general observations seem pertinent.

First, the relatively low levels of BI administered in combination with fasting therapy leave open the question of whether simply treating the patients with somewhat higher dosages of either LBI or UBI would provide results superior to those obtained with fasting plus BI. Fasting, after all, is not without cost in time, hardship, and compliance; and it can give rise to infections.

Second, the lack of a control group with ordinary drug therapy kept the researchers from forcefully underlining the superiority of their non-drug methods with hard statistical evidence.

Third, there was no long-term follow-up. From the American experience with the treatment of BA with UBI, it is clear that in many cases maintenance therapy is essential. New immune cells that can resuscitate both allergic and idiopathic BA tend to accumulate to troublesome levels within several months after the destruction of their predecessors by BI, so that perhaps four maintenance treatments per year are indicated, either permanently or at least for several years until the disorder ebbs away.

Fourth, the researchers appear to have known some way of dosing the LBI and UBI so that they were nearly equivalent. At least the fact that repeatedly the curves of the results crossed suggests that the dosages were roughly equivalent. Still, the lack of an objective metric for ensuring equivalence between the two kinds of BI leaves a fair degree of uncertainty. Since neither LBI nor UBI at the low dosages employed has any toxic side effects, it is very difficult to compare their strengths and toxicities. Needed is an objective means of measuring their comparative impacts on the body.

Fifth, the researchers' attribution of various causes for the comparative advantage of UBI over LBI seems unwarranted. Another explanation is that LBI has only a single wavelength whereas the Isolda UBI device and other UBI devices have at least a half-dozen scattered throughout the UV part of the spectrum and into the visible part as well. Thus the UBI might more closely resemble solar irradiation, to which the blood cells are presumably exquisitely attuned from primeval times; and so it might more profoundly stimulate the cells. The way in which

UBI forged past LBI in the second week of the fourth trial is very suggestive of the delayed action one would expect as a result of stimulation of hematopoiesis in the bone marrow. In effect, UBI would send a more intense message to the hematopoietic cells to produce just the right quantities and mix of new blood cells to bring order to the situation in the body. The implications of this hypothesis are examined in Appendix D.[101]

In Conclusion

While the lack of controls who received the standard drug therapy do not permit one to draw definitive conclusions regarding the value of BI in the treatment of bronchial asthma, the cumulative picture from the American and Russian studies is that BI--and especially UBI--appears to be an effective and safe treatment of this debilitating and widespread condition.

A closer inspection of the reports, moreover, reveals a highly interesting result. In effect, even though there were no formal controls, in reality there were *informal* controls. In the Russian trials, the fasting, fasting plus UBI, and fasting plus LBI groups served as controls for each other. All but the most rigid skeptics regarding the effectiveness of BI in general would have to concede that the pattern of UBI's superiority to LBI was unmistakeable, as was the pattern of their common superiority to fasting alone.

In the case of the 1946 American report, the main informal control was the grouping of the patients by age. In effect, each age group became a control for the others, as long as the goal of the study was taken to be the comparative effectiveness of UBI in the treatment of BA patients of different age groups instead of the formal one of demonstrating the effectiveness of UBI in the treatment of BA in general. The powerful conclusion that emerges from this informal but rather carefully controlled study-within-a-study is that UBI is far more

---

[101] A small Russian study of 10 patients with chronic obstructive bronchitis and 11 with BA yielded parallel results. After 5-9 treatments with UBI, in both groups 8 patients experienced "pronounced beneficial effects." Those who did not had exhibited the most severe complications at the outset (N.R. Paleyev et al., "The Efficacy of UBI in the Treatment of Chronic Nonspecific Lung Diseases [Russian]," Vestnik Rossiiskoi Akademii Meditsinskikh Nauk (1993), No. 3, pp. 3-6).

effective in the treatment of childhood BA than in the treatment of patients over 40. That still leaves open the question of whether UBI is more effective than competing drug therapies of BA, but it decisively answers the question of whether it had been demonstrated by means of a trial with controls that UBI was superior to any other treatment by means of restating the question. The new question is: Is any kind of UBI superior to any other treatment? The answer is that UBI as a treatment of childhood BA is decisively superior to UBI as a treatment of BA in adults over 40.

That still will not satisfy the skeptic, but it provides a very clear answer to the question of whether UBI was ever tested against BA by the American pioneer practitioners in controlled trials. In a formal sense, no. In an informal sense, yes. And the results were dramatic. In other words, these clinical studies were not of the sort that would win regulatory approval, but they contained scientific data and findings that deserve careful consideration by any serious investigator.

Closer inspection of the data in the 1943 BA study reveals another remarkable pattern as well. If one considers the results of the 9 cases out of 80 in which the patient was under the age of 20, a highly consistent pattern of "self-abnegating scoring" appears. This is a phenomenon encountered at times in studies conducted by researchers who are so eager to defend themselves against the anticipated charge of having slanted their findings to suit their purposes that they end up slanting them in the opposite direction. In 8 out of the 9 cases, the assessment of the result appears to be understated; in the ninth there was no follow-up. For instance, patient #22, a male of 14 with BA of 2 1/2 years' standing at the outset of UBI therapy, received a single treatment with UBI. Before that time, he had suffered from severe monthly attacks of BA. His condition was characterized at the end of the study 11 3/4s months later as "immediately relieved, no further attacks," but the result was scored as only "moderately improved."

This sample of 9 cases suggests that in fact this BA study--and perhaps the 1946 one as well--actually systematically *understated* the effectiveness of UBI.

A curious final note: in the standard textbook--Earle B. Weiss and Myron Stein, *Bronchial Asthma: Mechanisms and Therapeutics*[102]-- published at the same time as the report on the Vladivostok clinical trials and 50 years after the first American report, there is not a single word about the use of BI in 1,223 double-columned

---

[102] 3rd ed., Boston: Little, Brown, 1993

pages, not even in a chapter entitled "Inappropriate and Unusual Remedies" where dozens of unorthodox remedies are discussed. The closest that the authors come is to state "Occasionally other vaccines are still used by unscientific practitioners, including injections with the patient's blood (autohemotherapy)...."[103]

Why has BI become so invisible--as invisible as ultraviolet rays themselves? Two appear to be operating in this context. First, UBI certainly had its adherents in the United States, but it was never connected to any overall therapeutic approach with a long tradition and a permanent constituency. Only in Russia did UBI find such a framework in a country with a strong tradition in physics and a special interest in ultraviolet irradiation because of its long, dark winters. Second, an expert in any medical specialty could be forgiven if he or she would not immediately distinguish BI from the crowd of "unusual remedies." Unless there was some personal contact, the only reason such a specialist might have to investigate the BI literature or to try BI out in practice would be a compelling need to find a treatment for some harmful disease.

---

[103] p. 1148

Chapter 6: BI in Diseases of the Heart and Circulation

While BI's excellent effect in unblocking microcirculation appears to be a secondary operative mechanism explaining its overall efficacy in the treatment of respiratory diseases, this aspect of its functioning moves to the center of attention in the treatment of diseases of the heart and circulation, including cerebral circulation.

The early American UBI practitioners found that UBI worked well in peripheral atherosclerosis and against certain kinds of gangrene. They also did interesting work in applying UBI in various cardiological indications. It remained, however, to the Russians to employ UBI and LBI to cardiology in a thorough manner via a series of clinical trials and connected studies. Meanwhile, both Russians and East Germans thoroughly investigated the use of BI in the treatment of diseases of the lower extremities.

We will discuss first the cardiological indications, then the peripheral vascular diseases, and finally the applications of BI in cerebrocirculatory disorders--and by extension, in psychiatry.

Diseases of the Heart

Rheumatic fever was a major disease at the time UBI was first used in the United States. While it has since become uncommon in the US, the record that UBI practitioners compiled is significant as an example of BI's use in pediatric medicine.

In a study that took place in the late 1940s, 108 children with rheumatic fever were treated with UBI.[104] Of them, 22 were hospitalized with acute rheumatic carditis. 2 of these were in extremely severe condition at the outset of therapy, and one of these died. 9 had severe rheumatic carditis. 13 had severe to moderate rheumatic carditis. All except the single fatality were able to leave the hospital after UBI treatment without any sign of rheumatic activity. They and the 86 outpatients received prophylactic UBI maintenance therapy, with 8-10 treatments the first year, then 6 per year over several years. There were only 2 recurrences in the entire group of 107, and these were treated successfully with UBI. In contrast to treatments of other groups with sulfanilamide and other chemotherapies, there was not a single report of toxicity with the patients treated with UBI.

R.C. Olney used UBI to treat many patients with acute coronary occlusion and congestive heart failure. He never published his findings in a standard medical journal, but his informal observations and case studies have been recorded.[105] Olney considered that, in contrast to most other applications of UBI, it was most effectively used in heart disease as part of combination therapy. Rapid employment of UBI following acute coronary occlusion led improved microcirculation and reduced inflammation in the heart muscle. It also markedly relieved pain, reduced severe cyanosis and dyspnea, and when repeated every few days would even in cases that originally had appeared terminal lead to an overall improvement that would permit the patient to be discharged from the hospital. When patients failed to follow Olney's monthly UBI maintenance regime, their conditions would worsen and they would have to return to him for UBI treatment.

---

[104] V.P. Wasson, G.P. Miley, and P.M. Dunning, "Ultraviolet Blood Irradiation Therapy (Knott Technique) in Rheumatic Fever in Children," Experimental Medicine and Surgery 1 (1950), pp. 15-33

[105] Miley (1997), pp. 178-82

One major Russian study focussed on severe ischemic heart disease.[106] 70 males, 32 to 79 years old, with years of heart disease formed the study group. All patients had suffered from angina pectoris, and 56 had had a heart attack. All patients were receiving nitrate and other medications as well as nitroglycerine tablets. There was no control group.

The patients all failed an intensive combination drug therapy before receiving UBI. They then received on average 7 treatments with UBI, with the average of 1 ml per kg of weight--a level well below the Knott Technique range.

No toxic effects were noted. Following UBI therapy, almost all patients were able to reduce their intake of nitroglycerin, some to negligible levels. Instances of stenocardia were reduced. 46 patients succeeded in walking 1 km in the course of a day. 31 of 39 patients with jobs were able to return to work. All patients were followed for 2-16 months. 19 needed new UBI treatments, which yielded stabilized conditions. 7 patients died, of whom 6 were over 60 and 3 over 70. Three died from heart attacks.

The researchers assessed the results as "outstanding" in view of the serious condition of the patients. They did not observe any effect on the fibrogen level of the blood or on its fibrinolytic activity, nor on erythrocyte aggregation (these effects are often observed when larger dosages of UBI are used). They attributed the results to improvements in microcirculation, vasodilation, and oxygenation of the blood as well as to a general transformation of the blood as a response to changes in blood cell membranes as a result of the irradiation, including the dumping of various substances into the blood that had further beneficial effects. The effect of UBI on humoral immunity also lessened the post-infarct cardiosclerosis that can be caused by immune reactions.

In a study of stable angina pectoris, Ukrainian researchers compared different kinds of low-intensity laser therapy on 133 patients.[107] They had found that certain patients experienced an

---

[106] Irina E. Ganelina et al. "Zur Therapie schwerer Stenokardien mittels Ultravioletbestrahlung des Blutes (UVB) und zu einigen Wirkungsmechanismen dieser Therapie," Folia Haematologica 109 (1982), pp. 470-82

[107] V.M. Yurlov et al., "A Comparative Analysis of the Results of Using Various Forms of Helium-Neon Laser Therapy of Patients with Stable Stenocardia [Ukrainian]," Likarska Sprava (1995), No. 1-2, pp. 73-76

unexplained increase in distress symptoms after 4-5 treatments of local laser and LBI therapy, so they compared these modalities with low-intensity laser therapy of the liver projection area.

According to the researchers, aiming local laser therapy at the liver projection rather than at the heart or LBI yielded superior results. Their statistics appear to support this conclusion, which contradicts most other findings regarding the relative merits of local laser therapy vs. LBI. It would be interesting to run a similar study in which UBI would be compared with the other modalities.

Intracardial LBI

A third study investigated the effects of LBI during severe heart attacks.[108] 30 patients received 5-7 treatments for 40 minutes each. A control group of 20 patients received only drug therapy. The laser waveguide was led into the right forechamber, and the laser pulses were synchronized with the heartbeat.

The researchers noted a stabilizing effect on the patients' emotions, a powerful analgesic effect (in 33 percent pain was totally suppressed and in 22 percent it was significantly reduced); narcotics were reduced to 1/8 the level ordinarily used and analgesics to 1/3 the ordinary level. Within 2-3 hours of the start of LBI, the most intense pain level was felt by only 15 percent of the patients whereas 45 percent of the controls had this level of pain. The period following the heart attacks was also without complications for the LBI group.

There was a marked improvement in the rheological properties of the blood. Viscosity dropped by 30 percent, platelet aggregation by 25 percent, the level of fibrinogen by 20 percent--leading to a 35 percent reduction in general peripheral resistance and a normalization of diastolic pressure. The reduction in aggregation and the increased lability of red blood cells enhanced the improvement of microcirculation. There was a stabilization of hemodynamic parameters as well as a more rapid resolution of the heart attacks. The LBI group maintained these

---

108 A.A. Togaibaev and T.S. Alimzhanov, "Intravascular LBI in Combination Intensive Therapy of Patients with Myocardial Infarct [Russian]," Anesteziologiia i Reanimatologiia (1993), No. 5, pp. 45-47

benefits during a 6-month followup, whereas the control group underwent an increase in peripheral resistance and diastolic pressure.

According to the original Soviet regulations regarding the use of the Isolda UBI device, it was counter-indicated for use within three weeks of a heart attack. In addition, experiments suggested that UBI might exacerbate arrhythmias and lead to atrial fibrillation if used at the time of a heart attack. To test this, researchers in Kiev conducted a trial with 24 male patients within 6 hours of the start of a heart attack.[109] There were no controls. The patients received analgesics, sedatives, nitrates, calcium antagonists, a glucose-insulin-potassium mixture, and 3-5 sessions of UBI.

There were no side effects of UBI. 21 registered an analgesic effect. One patient with unstable stenocardia developed fibrillation as a heart attack began, and he died of progressive disruption of circulation.

Following UBI, incidents of extrasystole decreased sharply, then tended to increase after 12-24 hours, necessitating another treatment with UBI. Not only did no exacerbation of arrhythmia occur; to the contrary, UBI had an anti-arrhythmic effect, presumably because of its anti-ischemic action. The authors concluded that UBI was a useful component of the treatment of acute heart attacks. In other words, the regulations counterindicating its use within three weeks of a heart attack were contradicted by clinical results.

The Leningrad Studies

In the volume *Mechanisms of the Influence of Blood Irradiated with Ultraviolet Rays on the Organisms of Humans and Animals*[110], the researchers studied various aspects of the action of UBI in heart disease.

In one study, the authors noted that a series of reports had established the beneficial effects of UBI in the treatment of atherosclerosis of the heart and lower extremities. UBI improved the rheological characteristics of the blood and microcirculation, increased

---

[109] Y.N. Sirenko, I.E. Malinovskaiia, and S.S. Krasnitskii, "On the Treatment of Patients with Acute Coronary Insufficiencies with UBI [Russian]," Vrachebnoe Delo (1990), No. 10, pp. 9-11

[110] Ganelina and Samoilova (1986), *passim*

the oxygenation of the blood, and raised the level of cholesterol and beta-lipoproteins.

They were concerned, however, that the increase in lipid peroxidation caused by UBI might damage tissue in ischemic disease of the heart and lower limbs; and so they studied the impact UBI had on the pertinent enzymes. 67 individuals, including a control group of 13 healthy males and 54 patients, were studied. 25 had ischemic heart disease, and 29 ischemic disease of the lower limbs. 15 heart patients and 11 lower limb patients received UBI (5-10 daily sessions) as part of combination therapy. 11 heart patients and 6 lower limb patients were also operated on.

12 out of 15 heart patients and 8 out of 11 lower limb patients benefited substantially from UBI, the heart patients more than the others.

The authors concluded that UBI needed to be dosed with special care in the elderly, lest it activate lipid peroxidation in a manner that could be harmful to them. They found that the best results were in younger patients and those in earlier stages of disease, during the post-operational stage, and with inflammatory processes.

In another study, 145 males with severe ischemic heart disease following a history of heart attack over at least 3 years. The patients received 5-10 treatments with UBI along with standard drugs. There were no controls.

137 had a favorable response to UBI. Their overall conditions improved. They required fewer analgesics or nitroglycerin tablets. 92 of these had fewer incidents of stenocardia and were able to walk 1,000 meters per day around the ward. The other 45 saw moderate improvement. The researchers felt that UBI's regularization of biochemical substances and functions in the body accounted for the length of the maintenance of the good effects it brought about.

As a last representative of the dozens of other Russian and Ukrainian clinical trials of BI in the treatment of heart disease, an Izhevsk study merits notice. Researchers studied 45 patients with unstable stenocardia, of whom 26 had post-infarction cardiosclerosis.[111] A group of controls was selected that matched the study group in terms of age, gender, and severity of the fundamental illness.

5-7 courses of LBI were administered at 5 milliwatts for 30 minutes per session. All 45 patients underwent an improvement in general condition as well as a lessening of weakness, headaches, and

---

[111] Izhevsk (1994), pp. 38-39

insomnia. They all were able to have their nitrates reduced by 50 percent on average and gained a series of other benefits that were attributed to LBI.

Peripheral Vascular Diseases

These same effects of BI appear in the treatment of the range of disorders caused by vascular blockages in the legs and arms. The American pioneers used UBI to control the disease process in atherosclerosis in diabetes and other conditions, in some cases enabling them to avoid amputation of gangrenous toes and feet by healing the tissue under the surface gangrene, in other cases limiting the area that was necessary to amputate, and in still others reversing seemingly hopeless swelling and cyanosis.

In 1939 Miley treated one patient for thromboangiitis obliterans (Buerger's disease), a disease primarily of males between 20 and 40 years in age caused by cigarette smoking in which inflammation in small and mid-sized arteries in the extremities blocks the flow of blood. His account gives the flavor of the conditions under which the early UBI practitioners worked:

> Two UBI's were given 48 hours apart; a relief of pain occurred within two to three hours after the first UBI; after the second UBI it was decided by the surgeon to amputate the two gangrenous toes, which was done and the patient convalesced uneventfully from the operation. UBI was continued for about two years, with treatments one to three months apart. The patient was maintained in rather good condition and was able to work with no further progress of his thromboangiitis obliterans during this time. He then was told by other physicians that he had recovered spontaneously, probably did not have Buerger's disease, and UBI had no effect of any kind, might even be dangerous. He stopped treatment and soon afterward his thromboangiitis recurred with severe pain and rubor, etc. Again two UBI's were given, 48 hours apart, and again his pain disappeared. However, again he discontinued UBI completely, had a rapid recurrence of his Buerger's disease with

progressive gangrene, amputations and death in less than a year after discontinuing UBI.[112]

It is enlightening to compare Miley's initially successful treatment with the results obtained almost 60 years later in American treatments of thromboangiitis obliterans. In an effort to suppress pain, physicians must at times resort to continuous intravenous narcotics. They use a panoply of analgesics, antibiotics, vasodilating agents such as calcium channel blockers and alpha-adrenergic blockers (for which there are no controlled studies demonstrating efficacy in ischemic pain and wound healing), heparinization for lysis of clots (not totally satisfactory), prostaglandins for vasodilation and antiplatelet activity (unproven), corticosteroids (of little benefit), bypass surgery (effective in many cases but less so over time), sympathectomy (disappointing), and amputation.[113]

Thrombophlebitis

A disturbing sequela to surgery, childbirth, and other traumas is the development of acute thrombophlebitis in the legs, probably an autoimmune response of the body to perceived attack. In 13 consecutive cases treated over 4 years in the Blood Irradiation Clinic of Hahnemann Medical College and Hospital in Philadelphia (all with thrombophlebitis in the femoral and saphenous veins), the patients recovered nicely after 1 or 2 sessions of UBI. Five of the cases were failures of chemotherapy, and in one case 4 nerve blocks had been attempted. First pain and tenderness disappeared, then fever, and finally edema.[114]

---

[112] Miley (1997), p. 186

[113] Jess R. Young, Jeffrey W. Olin, and John R. Bartholomew, *Peripheral Vascular Diseases*. 2nd ed., St. Louis: Mosby, 1996, pp. 377-8

[114] G.P. Miley, "The Control of Acute Thrombophlebitis with Ultraviolet Blood Irradiation Therapy," American Journal of Surgery 60 (1943), pp. 354-60

The impressive results that BI can obtain in treating blockages in the arteries of the legs (arteriosclerosis obliterans) attracted a great deal of attention and clinical work in East Germany. Frick reported on a double-blinded clinical trial in which 50 patients in Fontaine Stage II of arterial disease were tested for their ability to walk following therapy. After 16 weeks of outpatient drug therapy, one group had improved by 160 percent. After 4-6 weeks of inpatient special treatment, a second group improved by 100 percent. A third group first received a 6-day "intrusive placebo" treatment with UBI in which the device was not actually working. The result was a 90 percent improvement. Then these patients received a genuine 6-day UBI treatment that led to a 360 percent improvement (270 percent over the placebo results).

This result was confirmed in 18 subsequent reports of similar trials.[115]

It should be noted that in the above trial, the results of placebo and treatment paralleled those found in other instances: BI frequently appears to have a significant placebo or psychological effect; and it has a pharmacological effect that is even stronger.

Frick noted that smoking by patients as well as diabetes made it much more difficult to obtain results and required many more treatments with UBI.

These results are markedly superior to standard pharmacological therapy. In the US, for instance, the only FDA-licensed drug for intermittent claudication (pain distal from the blockage) is pentoxifylline (Trental), but various clinical trials have shown that it provides only a 19-65 percent increase in walking distance. The only experimental treatment that can match the results of UBI as reported by Frick is prostaglandin E_1, and that is likely to have side effects that are considerably more toxic than those of UBI.[116]

---

115 Frick (1989), pp. 56-57

116 Young et al. (1996), p. 223

In a study of LBI monotherapy in the treatment of ischemic disease of the lower limbs with the avowed goal of distinguishing its action from those that might occur in combination therapy of which LBI was just one part, Russian researchers undertook a trial with 27 patients.[117] The patients received 5-6 treatments of 30 minutes of LBI at 5 milliwatts. 19 of the patients were able to walk without pain an average of 350 percent farther. The other 8 patients experienced a stabilization of ischemia only. The lack of a control group and of a clear baseline detracts from the value of this trial, but it remains of interest because it was one of the few in which the Russians have used BI in cardiovascular disorders outside of the framework of combination therapy.

Lymphedema, the accumulation of lymph ordinarily in the tissues of the feet and legs, is a painful condition that can lead to ulceration and further complications. While it can be treated conservatively, the outcome is often disappointing. Russian doctors therefore ran a trial of UBI on 48 female patients, with a control group of 19 healthy women.[118] 34 of the patients had frequent erisypelatous skin--21 at the time of entry into the clinic.

The patients received 5 treatments of UBI. While all of them felt a lessening of pressure and pain following treatment, the best effects were felt by those with erisypelatous conditions. Their temperatures dropped, and the exudate was reduced. 6-8 months of follow-up showed that none of the patients required more UBI. 30 of the 34 who had had erisypelas prior to treatment showed no more signs of it, and 3 more had only minor signs of it. The researchers particularly recommended UBI for cases that would require operations as a pre- and postoperation treatment that greatly simplified and improved the operation itself.

A review article surveying the Russian experience in LBI treatment of obliterating atherosclerosis and endarteritis of the lower extremities found that there was no evidence at all of damage to arteries caused by laser irradiation even in experiments in which the laser was used two hours per day for ten days. The figures for improvement following LBI were consistently in the range of 70-75 percent of patients across several studies. Armenian researchers even reported regression

[117] Brill' (1994), pp. 184-86

[118] Ganelina and Samoilova (1986), pp. 119-21

of plaques and revascularization of formerly blocked arteries that occurred beginning with the second day following LBI and persisted.[119]

It is correct to be skeptical of such claims. Were they true, then they would presumably have been reported by many other Russian and Ukrainian specialists, not to speak of the East German and the pioneering Americans. One possible explanation is that indeed BI can temporarily clear up the flow of blood through blocked arteries and veins as long as the damage is not too severe and longstanding. Once unblocking occurs, then the flow of blood, access to oxygen and to nutrients, and improved diet and exercise can in combination over time allow tissue to regenerate and plaques to be sloughed off. This can occur anyway without BI or other interventions in the case of plaques in the coronary artery, for instance, in perhaps 1 out of 3 cases. (Of necessity, of course, these are cases in which there was no complete blockage.) BI's role would be to initiate and foster this process, but it is unlikely that BI alone would suffice.

Cerebral Vascular Disorders

BI appears to encounter no difficulties in passing the "blood-brain barrier," in contrast to certain drugs, though the mechanism by which this occurs is not clear. It has the advantage over local irradiation of the brain of being bloodborne and thus less likely to damage delicate tissue. Given BI's excellent track record in the treatment of vascular disorders elsewhere in the body, it is not surprising that physicians have studied its effects on cerebral vascular disorders as well.

An East German study compared the effects of UBI (15 patients) with a one control group of 15 that received a thrombolytic, antihypertensives, glycosides, and physiotherapy similar to those received by the UBI group, and a second control group of 15 healthy volunteers. The patients had been diagnosed as having cerebrovascular insufficiency, stage IIIb-IV. UBI was administered in 7 sessions.

The UBI group outperformed the first control group on 24 of 26 variables, and it showed considerably greater improvement (96 percent) over the period of the study than did the second control group (73

[119] Brill' (1994), pp. 138-44

percent), of which the improvement could only be attributed to a practice effect when taking the same tests a second time. The UBI group did especially well relative to the first control group in cognitive speed, higher perceptual processes, verbal recognition, and concentration.[120]

An important Russian study of UBI treatment of neurological disorders involved 90 patients aged 47-69 with atherosclerotic, hypertonic, and venous circulatory dysfunction refractory to other treatments or gaining only short remissions with them.[121] There were 35 controls. 4-8 UBI treatments were given. Positive results were obtained with 87 percent of patients, including a full resolution of 51.2 percent of the neurological symptoms of the 37 atherosclerotic patients. BI treatment caused the disappearance or significant decrease of headaches, dizziness, tinnitus, feeling of heaviness in the head, pain in the heart region, etc. Sleep was normalized as well. In most positive cases, the results were longlasting or permanent.

A team of Odessa physicians studied the effects of UBI used as a monotherapy on the vegetative neural system of 50 sailors, aged 40-60, with early-stage cerebral circulatory problems. They used a Photon UBI apparatus and treated 0.5-1 ml per kg of each patient's blood 5 times in the course of a single day.

90 percent of the patients experienced subjective improvement. Their heads cleared, a feeling of weight on their heads disappeared, headaches diminished, tinnitus ceased, they felt more ready to work, their mood improved, and their sleep normalized. The lack of controls and followup diminish the value of this study, but it is still of some interest as one of the few Ukrainian or Russian studies of BI as a monotherapy.[122]

---

[120] Dagmar Klink et al. "Ultraviolett-Bestrahlung des Eigenblutes (UVB) bei zerebrovaskulaerer Insuffizienz unter besonderer Beruecksichtigung der Hirnleistung (Pilotstudie)," Zeitschrift fuer Klinische Medizin 42 (1987), pp. 1145-49

[121] M. Ia. Berdichevskii and E.M. Dashkovskaia, "Effectiveness of the Complex Treatment of Cerebrovascular Disorders by Ultraviolet Irradiation of Autologous Blood [Russian]," Zhurnal Nevropatologicheskoi Psikhiatrii Im. S.S. Korsakova 91 (1991), No. 1, pp. 75-78

[122] A.A. Lobenko et al. "Vegetative Mechanisms of Initial Appearances of Insufficiency of Cerebral Blood Circulation and a Method of Correcting Them [Russian]," Likarska Sprava (1993), No. 1, pp. 77-79

Schizophrenia

BI's effectiveness in treating cerebral vascular disorders can also be seen in two highly interesting studies of its use in the treatment of schizophrenics resistant to all other therapies.

At an institute for neurology and psychiatry in Kharkov, 62 patients were diagnosed as having simple or paranoid schizophrenia ranging from 1 to 23 years in duration. They exhibited a wide range of symptoms, from hallucinations to apathy. They all had progressive syndromes resistant to various therapies including psychotropic drugs, sulfazine, insulin comas, and Electroconvulsive Therapy (though it is not clear whether all of them had been treated with ECT). The patients received 8-15 LBI treatments on a daily basis. The researchers used only 0.2-1.0 milliwatt doses for 1-2 minutes and remarked that this was much lower than the amount ordinarily used. They justified this approach on the grounds that the schizophrenic patients' nervous systems were much more sensitive to LBI.

21 out of 38 patients (55 percent) with remitting schizophrenia responded well to LBI. Among them, the therapy had its best results in the depressive-paranoid group (11 out of 14), but was less effective with the hallucinatory-delirious group (8 of 19). In this second group, those who improved experienced a diminution in delirious ideas, a reduction in hallucinations, and a lessening of motor and ideational monotonous behavior. Their moods gradually improved. Interest in their surroundings appeared. Their social contact and desire to join in strolling around and in work improved. Their sleep normalized. 5 attained remission with full grasp of their condition. Upon returning to normal life, they had superior ability to adapt and behavioral adequacy. Some of them returned to work.

The hypochondriacs experienced no benefit from LBI, nor did those with a severe Kandinsky-Clerambeau syndrome.

Of the 24 patients with non-remitting, progressive schizophrenia, only 9 benefited (37.5 percent), and their improvement paralleled that of the schizophrenics in the other group.

The researchers noted that with patients in whom parasympathetic tonus of the vegetative nervous system predominated, it was possible to use LBI as a monotherapy. But with patients in whom tonus of the sympathetic nervous system predominated, LBI should be used only in combination with intensive pharmacotherapy. Otherwise, it might provoke neuralgia, increased heartbeat, increased arterial pressure, heightened anxiety, and weakness. Four patients in the remitting group experienced these symptoms after 2-5 treatments with LBI, but they quickly disappeared when drug dosage was boosted.[123]

What conclusions can be drawn from this report? The lack of a control group made it difficult to judge whether the improvements noted were in fact to some extent the result of a placebo effect or simply represented temporary remissions in the cases of the patients with remitting schizophrenia. Still, LBI did appear to have enough physical effect to cause physical trouble for four patients. One must ask whether the patients might have experienced even better results had the treatment been with UBI and had it been at a somewhat higher dose. The argument that the patients' nerves were unusually sensitive to LBI does not appear convincing in light of the many other cases in which inadequate doses of BI led to inadequate effects. It should be noted that both the researchers and their editors were cautious about the use of what was a new therapeutic approach in psychiatry.

Still, quite a few patients whom nothing else had helped seemed to benefit significantly from the treatment. Presumably in these cases BI's capabilities of relaxing the autonomic nervous system, oxygenating the blood, dilating blood vessels, disintegrating blockages of microcirculation, and improving the rheological properties of the blood were the main modes whereby it achieved its effects.

A second study examined the use of LBI in the treatment of schizophrenic patients with a depressive syndrome.[124] 25 patients with an average of 9.3 years of diagnosed schizophrenia and over 6 months of depression were treated; all had developed resistance to antidepressants and other psychotropic medications. LBI was administered in a dose of not less than 1 milliwatt for 30 minutes per day over 10-15 days. The patients continued to receive their standard drugs, to which was added "Aevit", in order to suppress any (unspecified) side effects of LBI.

To measure the effects of LBI on cerebral circulation, the researchers used doppler sonography and telethermography (to measure infrared emanations) of the carotid artery on 8 patients.

---

[123] I.I. Kutko et al. (1992), pp. 53-56. This might have been one of those instances in which the dosage was so low as to exacerbate the problem in four patients rather than solving it.

[124] I.D. Stulin et al., "Dynamics of Cerebral Circulation in the Treatment of Schizophrenia Patients with a Depressive Syndrome by the LBI Method [Russian]," Zhurnal Nevropatologii i Psikhiatrii Imeni S.S. Korsakova 94 (1994), No. 1, pp. 54-56

After 3-5 treatments, sleep and appetite became normal. There was more motor and mimetic activity as well as more contact with the surroundings. After 8-10 treatments, mood became more equilibrated. Psychic anesthesia, inferiority complex, and suicidal tendency disappeared. Depressive fantasies lost their strength. In contrast, increased activity and readiness to work came much more slowly, and as a delayed effect after the end of the therapy. In those patients with adynamic depression marked by ideational disturbance, the recuperation took longest. Even when their activity and ability to work had improved, they remained distrustful of their intellectual capabilities, and memory and attention deficits persisted. Still, the researchers considered that all 25 patients entered into remission from depression following LBI.

Doppler sonography and telethermography on 6 patients revealed that previous circulatory deficits had virtually disappeared in 5 of them.

There were no concurrent controls, follow-up, or indications of how remission was measured. Still, the results of LBI appear to have been highly satisfactory in consideration of the patients' resistance to other therapies, which in this case served as a kind of control. In effect, the study suggested that a certain percentage of cases of depression arose from cerebral circulatory disorders, and that LBI was an excellent remedy.

Hypertension

High blood pressure is, of course, a major source and symptom of cardiovascular disorders. BI can have significant effects in reducing it.

In a trial held in Baku, Azerbaijani researchers used an extracorporeal low-intensity laser device to irradiate blood before reinfusing. 73 Stage 2 hypertensive patients were divided into a group of 24 who were treated with standard drugs and a group of 39 who received standard drugs plus the laser therapy. This second group was characterized by severe hypertension that was resistant to drugs. 5-7 treatments of 10-15 minutes were carried out with a single day's interval between them.

Blood pressure dropped 24 percent in those treated with laser therapy. The patients experienced a lessening of headaches and dizziness as well as a diminution of aching in the heart region. An increased responsiveness to drugs permitted the doses of drugs to be reduced on average by 40 percent. After leaving the hospital, the patients had remissions lasting 4-8 months whereas the patients receiving only drugs had remissions that did not surpass 4 months. In the laser-treated patients, the reduction in hypertension occurred 6-10 days after start of therapy, whereas with the drug-only patients it took 14 days to occur.[125]

In East Germany a study was done of 14 hypertensive patients over 3 years.[126] Treatment with UBI in addition to hypertensives reduced initial systolic blood pressure from 175 on average down to 148 and held it below 153 for 3 years. It reduced diastolic pressure from 94 to 86 in 6 weeks and then to 83 in two years, but in the third year it rose to 91. In addition, it was found that maintenance treatment with UBI pushed the normalized blood pressure of two patients into the low blood pressure area, and so UBI was no longer given to them. The general conclusion was that UBI significantly reduced blood pressure in hypertensive patients. In 25 percent it was possible to reduce antihypertensive medications by 1/3. The lack of a control group deprived this study of greater value, but it still provides some insight into the benefit that treatment with BI might provide.

In Conclusion

UBI and especially LBI have become standard parts of combination therapy in Russia and Ukraine for a range of acute and chronic heart conditions. All of the evidence in clinical trials reported suggests that BI provides many benefits and does not cause the problems that were initially of concern such as fibrillation. BI's vasodilatory effect, its ability to unblock microcirculation, its relaxation of muscle via its action on the autonomic nervous system, its anti-inflammatory action, its anti-microbial action, and its oxygenation effect make it a very helpful tool in general cardiology.

---

[125] I.G. Alizade and N.T. Karaeva, "An Experiment in the Use of Autotransfused, Laser-irradiated Blood in the Treatment of Patients with Hypertensive Disease [Russian]," Likarska Sprava (1994), No. 5-6, pp. 29-32

[126] Frick (1989), pp. 62-64

American, German, and Russian practitioners who have used BI in peripheral vascular disorders are unanimous in noting its effectiveness even in many difficult cases that yield to no other intervention short of amputation. If amputation becomes necessary, BI can often save the most tissue possible. Its analgesic property is also a great advantage in what can otherwise be very painful conditions. BI has not been thoroughly tested in various other vascular disorders, though individual practitioners have treated Raynaud's syndrome, varicose veins (BI reduces accompanying inflammation and is prophylactic against infection), lymphangitis, and frostbite.

In the treatment of cerebrovascular disorders, BI can have powerful action. In turn, this can enable it to treat certain neurological and psychiatric syndromes with greater effectiveness and lower toxicity than drug therapy. An important question is whether BI is effective in patients resistant to Electroconvulsive Therapy. The above studies suggest that in fact it can be very effective in this indication, which in turn raises the question of whether BI is actually a humane substitute for ECT in every case where ECT is currently used.

Chapter 7: BI in Gastrointestinal and Kidney Diseases

Certain disorders of the body's central organs have yielded to the advance of medical science; others remain beyond its reach. The most spectacular success in recent years is the discovery that peptic ulcers are mainly caused by bacterial infection with Heliobacter pylori, against the prevalent view that a bacterial etiology was unthinkable. This has permitted pharmaceutical companies to come up with a series of very effective treatments for ulcers. In fact, all along BI had been an effective treatment, but few were paying attention. The pioneering American UBI practitioners used UBI against ulcers, and the Russians took the trouble of conducting trials of BI against them. But there is little call for employing BI in this indication any more because the drug therapies are almost totally effective and are even cheaper than BI.

In contrast, chronic hepatitis B and C remain serious and largely untreatable. Interferon has proven successful in a fair proportion of cases; but many patients experience a resurgence of disease within months, and the toxicity of interferon is quite high. New antivirals are expected to be more effective and less toxic, but even they are not likely to represent a fully satisfactory treatment of these difficult conditions.

The key study of the use of UBI in viral hepatitis, that of Olney and his collaborators in 1955, was already discussed in Chapter 4.[127] Here it is appropriate, however, to make several comments about the study and its implications:

- as conclusive evidence, the study is lacking because there were no controls and because it was in effect a meta-analysis of sorts drawn from the cases of 16 practitioners without the full, prospective

[127] Olney (1955)

framework that would give it greater credence. Still, the evidence is powerfully suggestive that UBI was effective in suppressing both acute and chronic hepatitis;

- as the researchers comment, UBI's ability to prevent recurrences is very important. Even in the treatment of acute hepatitis, ordinarily self-limiting, this can be of value because it prevents the 5-10 percent or so of acute infections from turning into chronic ones; and
- the researchers compared UBI as employed in their study with statistics from the use of other therapeutic interventions, with UBI superior in terms of duration of illness (19.2 days vs. 42-56 days), recurrences (4.65 (in an IV drug user) vs. 0.6-18 percent), and deaths (0 vs. 0.2-19 percent).

The effectiveness of BI in the treatment of chronic HBV and HCV also may imply that it would be equally effective in the treatment of autoimmune hepatitis (chronic active hepatitis). Its chief therapeutic competitor in this indication is glucocorticoids, which of course have significant toxic effects when used in long-term immunosuppressive therapy.

It should be noted that there has been no reported instance of the use of BI in the treatment of Idiopathic Thrombocytopenic Purpura (ITP), an autoimmune disorder of the spleen. Given the prevalence of ITP and the expense and shortcomings in existing treatments, BI would appear to be a very promising candidate, especially because the blood filtering action of the spleen would tend to concentrate the secondary radiation of BI as it does in the liver in hepatitis.

Biliary Disease

R.C. Olney was the first to report on his findings in regard to the treatment of biliary disease with UBI.[128] He treated 383 patients: 3 almost moribund; 5 very severe; 264 chronic, without stones; 56 chronic, with stones; and 55 chronic cholangitis and hepatitis, gallbladder previously removed.

---

[128] Olney (1946), pp. 235-7

Almost moribund patients aged 78 and 80 responded rapidly to UBI and left the hospital in 14 and 18 days, respectively, without surgery, Olney reported. They were in good health one year later. The third almost moribund elderly patient had a decompensated heart and so made slower progress after UBI treatment; but she, too, eventually recuperated and was reasonably healthy after one year.

4 of the 5 severely ill patients recovered after UBI without an operation, and the fifth did so after an operation. Of the third group of 264, all but 12 achieved "marked relief of symptoms and have been observed for a period of four months to one and a half years following treatment," he noted. Of the fourth group (chronic with stones), 48 of the 56 were operated on successfully with UBI as pre- and postoperative prophylaxis; Olney did not report on the other 8. 46 of the 55 in the last group were "greatly relieved of symptoms" and remained symptom-free for 4 1/2 months to 1 1/2 years of follow-up. The others in this group showed moderate improvement (undefined).

Olney noted that UBI had excellent effects in preventing or limiting peritonitis, ileus, pain, pulmonary complications, and phlebitis. He saw the primary action of UBI as the reduction of swelling and pressure caused by inflammation in the biliary duct and nearby areas.

It is worth noting that the current treatment of choice for gallbladder diseases is the moderately invasive operation of laparoscopic cholecystectomy. Olney succeeded in treating at least some of the same conditions with drainage and UBI alone.

In 1950 E.W. Rebbeck reported on the use of UBI in biliary tract surgery. He treated a series of 110 consecutive patients pre- and postoperatively with UBI and compared them to 226 other patients in the same series.[129] Rebbeck administered one UBI treatment on the day before the operation and 1-6 starting on the day following the operation and subsequently as indicated, the average patient receiving 2.

Excessive nausea and vomiting necessitating the use of gastric intubation occurred in 33.1 percent of the controls and only 2.7 percent of those treated with UBI. 28.8 percent of the controls had excessive abdominal distension as measured by the amount of usages of stimulants, laxatives, etc. as compared to 11.8 percent of the UBI group. Temperatures above 102° F were found in 32.3 percent of the controls and in 15.4 percent of the UBI group. Thrombophlebitis (phlebothrombosis) was diagnosed in 0.9 percent of the controls and in

---

[129] E.W. Rebbeck, "Use of Ultraviolet Blood Irradiation (Knott Technic) in Biliary Tract Surgery," American Journal of Surgery (1950), pp. 108-12

none of the treatment group. Mortality in the controls was 2.2 percent; in the UBI group it was 0.9 percent (one 87-year old woman who had had thrombophlebitis for 3 weeks prior to the operation subsequently developed fatal pulmonary embolus after fully recovering from the operation). Penicillin was used less frequently and in smaller amounts in the UBI group.

In other words, the pre- and postoperative use of UBI yielded very beneficial results.

A Russian study of the use of various types of "quantum" therapy in the post-operational treatment of elderly patients with cholecystitis arrived at comparable conclusions.[130] A control group of 16 received standard drug therapy. A second group of 20 received extracorporeal x-ray irradiation of the blood. A third group of 20 patients received standard UBI. A fourth group of 19 received donor blood that had been irradiated with ultraviolet irradiation from a device of the treatment center's own construction.

The researchers found that the three "quantum" methods were superior to the standard therapy, and that the fourth group fared better than the other two "quantum" groups. However, their failure to provide an objective basis for the comparison of the doses of the various irradiation therapies as well as a means of distinguishing between the irradiation and the "donor" components in the treatment of the fourth group undermined the value of the study.

There is no logical reason why donor blood would be more effective than a patient's own blood, and the expense and possibility of complications would appear to argue against this approach without such a reason being provided. Still, a head-on-head comparison of donor blood vs. a patient's own blood with otherwise identical BI therapy might prove interesting, even if the explanation for whatever differences would appear might not be easy to identify.

Another Russian study included 45 healthy controls and 130 patients with acute cholecystitis, of whom 85 received LBI and 40 standard medications. In addition, 65 of the patients received local magnet therapy. The researchers found that LBI was significantly

---

[130] R.B. Mumladze et al., "Comparison of Various Methods of Quantum Hemotherapy in the Treatment of Complicated Forms of Acute Cholecystitis in Elderly and Very Old Patients [Russian]," Vestnik Khirurgii Imeni I.I. Grekova 152 (1994), No. 1-2, pp. 112-5

superior to standard therapy, and that the addition of magnet therapy increased the beneficial effect.[131]

Pancreatitis

Acute pancreatitis is a dangerous disorder that can be very hard to treat, though many cases are self-limiting. When it enters into the necrotizing phase, it can be fatal. Russian researchers used UV-irradiated donor plasma in the combination therapy of 14 cases of necrotizing pancreatitis.[132] 10 of the patients had hemorrhagic pancreonecrosis; 8 were in serious condition with pronounced symptoms of enzymatic toxemia. There were no controls.

The patients received a barrage of therapies, including medications, local hypothermia, fasting, and various surgical interventions. When these proved inadequate, 1 ml per kg of body weight of donor plasma was irradiated on a UV device of local provenance (253.6 nm).

The authors found that the level of enzymatic toxemia (e.g., the level of tripsin in the blood) dropped by more than 2 times to near normal levels 5-6 hours after infusion of the irradiated plasma (from 1,213.4 to 635.5 ng (ml)). Insulin resistance quickly declined, and other indicators similarly improved. 3 patients died, all after operations for purulence in the pancreas. The researchers judged that a decisive factor in mortality was the late hospitalization of the patients, though even the 3 who died had experienced improvements in various indicators following UBI.

In a second study of acute pancreatitis, 65 patients were treated with a total of 83 sessions of UBI before, during, or after operations.[133] Patients who did not undergo operations were also treated with UBI. The researchers found that after UBI the patients' condition improved, appetite returned, tachycardia lessened, fever declined, and laboratory results normalized. There was evidence of immunostimulation but no

---

[131] Sukhanova (1993), pp. 132-8

[132] I.N. Markov, S.M. Chudnykh, and O.E. Kolesova, "Use of Donor Plasma Irradiated with UV in the Therapy of Destructive Pancreatitis [Russian]," Khirurgiia (1994), No. 3, pp. 28-29

[133] V.I. Filin, V.I. Koval'chuk, and V.N. Kravets, "UBI of Patients with Acute Pancreatitis [Russian]," Klinicheskaia Khirurgiia 11 (1984), pp. 28-29

side effects from UBI. Still, 11 of the 40 patients with purulent-necrotic pancreatitis died. The lack of controls prevented this study from having greater validity. More important, the inclusion of UBI into combination therapy and the very low average number of UBI treatments meant that UBI was not being used nearly as effectively as it could have been.

In a third study, researchers treated 60 patients, of whom 47 had chronic pancreatitis and 13 acute pancreatitis. The patients received 5-7 treatments of LBI (632.8 nm) at the rather high level of 5-7 milliwatts for 30-40 minutes as part of combination therapy following the failure of standard drugs. 92 percent experienced a disappearance of pain and vomiting, 83 percent had a lessening of nausea, 87 percent had improved appetite, and in 83 percent there was a reduction in the distension of the belly. As an indicator of improved functioning of the pancreas, in the acute cases the level of amylase in urine dropped from 1826.82 +/- 401.4 to 52.77 +/- 4.9 g/l ($p<.05$).[134]

How well might BI perform if used optimally in necrotizing pancreatitis? Perhaps the best way to respond is to say that, if used early on and in a high enough dose as a monotherapy--or in conjunction with physical methods such as drainage--BI has the potential to suppress the disease process and keep it from ever entering a necrotizing phase. Even in the necrotizing phase, treatment with perhaps 8-10 sessions of UBI might prove superior to all other existing measures. The statistical results on inactivation of trypsin in the first Russian study are much better than those obtained with most competing therapies. The American pioneer UBI practitioners reported treating many cases of acute pancreatitis successfully. UBI suppressed inflammation, relaxed the sphincter of Oddi (effects on the autonomic nervous system), and returned amylase and lipase values to normal.[135]

Appendicitis

While no clinical studies have been done, Miley reports that UBI can be used as a substitute for surgery in cases of acute appendicitis. Typically, he would administer UBI as a preparation for surgery, then wait, with the patient kept ready for operation. Almost invariably, he

[134] Izhevsk (1994), pp. 16-18
[135] Miley (1997), pp. 159-60

says, all signs of acute appendicitis would disappear. He followed one case for ten years, and there was no recurrence. However, for more complicated cases in which there were recurring symptoms and an appendiceal abscess, an operation is probably necessary. Miley makes it clear that neither he nor other UBI practitioners would recommend dispensing with surgery in order to treat appendicitis with UBI. He simply suggests that the patient be given UBI and prepared for surgery, then the surgeon should wait to see whether symptoms subside so that surgery becomes unnecessary.[136]

Peritonitis

Many of the early American UBI practitioners were surgeons, and so they frequently encountered peritonitis. In a 1942 study of 72 patients (without controls), of whom 29 had failed sulfa therapy, three subgroups were treated: 40 with generalized peritonitis; 20 with appendiceal abscesses plus localized peritonitis; and 12 females with multiple pelvic abscesses (Pelvic Inflammatory Disease) and severe pelvic peritonitis. 43 had moderately advanced peritonitis; they all recovered following UBI treatment. 29 were apparently moribund. Of these, 9 out of 17 in the first group recovered, 4 out of 7 in the second group recovered, and 6 out of 9 in the third group recovered. The other moribund patients died, two of sigmoid carcinoma.

The physicians noted that UBI treatment rapidly resolved paralytic ileus and led to rapid detoxification in the successful cases.[137]

In a recent Russian study, 35 patients with disseminated peritonitis of varying etiologies found that the inclusion of UBI in combination therapy reduced mortality to 4 out of 35 in comparison with 10 out of 37 patients for controls treated with combination therapy without UBI. The UBI group experienced a sharp increase in the numbers of T-cells (60 percent) as well as a concomitant decrease in

---

[136] Ibid., pp. 154-5

[137] G.P. Miley and E.W. Rebbeck, "Ultraviolet Blood Irradiation in Peritonitis," Review of Gastroenterology 10 (1943), No. 1, pp. 1-26; reprinted in Miley (1997)

circulating immunocomplexes (36.5 percent). UBI patients received 109 sessions altogether, or roughly 3 each.[138]

BI does not appear to have been surpassed by antibiotics in the treatment of severe peritonitis. According to a recent commentary,

> The prognosis for severe peritonitis has not changed in the past 50 years despite the advent of broad-spectrum antibiotics, critical care units, and radical approaches to eliminate bacterial contamination of the peritoneal cavity.[139]

Inflammatory Bowel Disease

Various physicians have used BI on ulcerative colitis without much success. There are no reports of its use in Crohn's Disease. In both cases, BI might be effective in early stages of the disease.

Kidney Disease

Anecdotal reports from the early American practitioners state that they had success in treating acute glomerulonephritis, but that follow-up treatment was needed to ensure that the symptoms did not return. Similarly, one 65-year old male patient with chronic glomerulonephritis and nephritis responded to UBI with a gradual, complete resolution of symptoms. He did, however, require monthly or bimonthly maintenance therapy. The outcome of treatment of nephrotic syndrome with UBI varied from full recovery to a downward trend until death. BI is not helpful in obstructive uropathy. Pyelonephritis is said to respond well to UBI, though in chronic cases maintenance therapy is required.[140]

---

[138] B.M. Ashurov et al., "UBI in the Treatment of Disseminated Peritonitis [Russian]," Khirurgiia (1997), No. 4, pp. 44-47

[139] J. Claude Bennett, ed. *Cecil Textbook of Medicine*. 20th ed. Philadelphia: W.B. Saunders, 1996, p. 746

[140] Miley (1997), pp. 167-72

In a Russian trial, 12 patients with chronic glomerulonephritis without disruption of kidney function were treated with LBI. The dose was 2-3 milliwatts for 7-10 sessions of 30 minutes. The results were favorable according to many indicators. Proteinuria, for instance, dropped from 1.34 to 0.71. In addition, the 7 patients with hypertonic disease experienced a reduction in systolic blood pressure from 180 to 145.2 mm and diastolic blood pressure from 118.4 to 88.5 mm.[141]

Three Siberian researchers conducted a clinical trial of local laser therapy and LBI in the treatment of chronic pyelonephritis with 61 patients, of whom 67.4 percent had urolithiasis and 32.6 percent adenoma of the prostate.[142] A first group of 17 patients received standard antibacterial therapy, which was fully successful in 20 percent of them. A second group of 11 received local laser therapy in addition to standard therapy; it was fully successful in 57.1 percent. A third group of 33 received LBI in addition to standard therapy, with full success in 64.3 percent.

The third group had the best results of the three in terms of normalization of both cellular and humoral immunity as well as speediest recovery; the second group had good results, but its cellular immunity remained considerably in deficit.

One weakness in comparisons of this sort is that the addition of almost any decent intervention to a standard combination therapy can yield a superior result simply because it ups the overall dose of medicine (though the opposite can also occur). So this trial's main finding was not that the two types of laser therapy added on to combination drug therapy were superior to combination drug therapy alone but rather that LBI was superior to local laser therapy. The researchers point out why this should be so:

> ...the efficacy of laser therapy depends not only on its local action but also on its way of provoking a stimulation of various systems and organs. LBI shows bactericidal action, activates the metabolism of substances, and improves microcirculation and the rheological properties of the blood. It leads to the removal of all hypoxia caused by whatever pathological condition or process; it

---

141 Izhevsk (1994), p. 44

142 A.I. Neimark, Z.T. Malazoniia, and E.B. Karabasova, "The Capabilities of Local and Intravascular Laser Irradiation for Removing Immune Deficiencies in Patients with Chronic Pyelonephritis [Russian]," Urologiia i Nefrologiia (1995), No. 2, pp. 27-29

effects the release in the organism of a cascade of the patient's own central and peripheral autoregulating systems of adaptation, which medicinal substances do not do.

In other words, it should not surprise us that LBI (and UBI) have superior overall action to local laser therapy not only in kidney conditions but in a wide variety of disorders. Curiously, it might also be possible to interpret the superior rate of clearing of bacteria by LBI in comparison with local laser therapy in this case to a stronger local effect as well, although this local effect is actually achieved by LBI's capability of summoning to the scene of the problem the activated secondary radiation of the blood cells.

Conclusion

This is hardly a comprehensive listing of all the indications in the treatment of these disorders of internal organs that BI has been employed against or could presumably perform well against. Instead, it illustrates the ways in which BI can be used as well as the limits it encounters. The overall picture is one of a highly versatile and effective therapeutic approach.

Chapter 8: BI in Gynecology and Obstetrics

The first two patients treated with UBI, in 1928 and 1933, were women moribund with septicemia after septic abortions. Both recovered uneventfully. BI has since been used in many cases of septic abortion with considerable success. This is a more difficult indication than ordinary septicemia because of the decline of immunity owing to pregnancy, the loss of blood, intravascular hemolysis leading to shock, the action of the toxic substance used to induce abortion, and liver and kidney malfunctioning.

One Russian study of septic abortions in which UBI was combined with hemoabsorption found a fourfold decline in mortality and a more rapid recovery on average.[143] Another studied 126 patients with sepsis, of whom 66 had undergone septic abortions outside the hospital. Three groups of 42 each were formed. Group 1 received 3-10 sessions of UBI. Group 2 received 3-10 sessions of UBI plus 2-6 sessions of hemoabsorption. Group 3 received standard drug treatment. A control group consisted of 35 normally pregnant women.

In Group 1 there were 2 cases of septicopyemia and 14 deaths; in Group 2 there were 0 cases of septicopyemia and 6 deaths; and in Group 3 there were 6 cases of septicopyemia and 23 deaths.[144]

It would be useful to have further details regarding the composition of the three treatment groups. Were they carefully randomized? Were they comparable? Were the average numbers of UBI treatments given to Groups 1 and 2 roughly the same? What exact process did the "hemoabsorption" procedure entail?

---

[143] A.I. Maltsuyev et al. "The Use of UBI in Obstetrical-Gynecological Practice [Russian]," Akusherstvo i Ginekologiia (1990), No. 8, p. 8

[144] Ganelina and Samoilova (1986), pp. 31-38

If the answers to these questions show that the trial was run in an unbiased fashion, then this would be one of the few instances in which a clear advantage accrued to treating patients with a combination of BI and some other therapy over BI alone. In that case, the question would be whether administering a higher level of BI that would equal the hemoabsorption therapy in toxicity would provide an equivalent effect--or a better one. At any rate, there is an interesting suggestion here that BI might be more effective if used in tandem with hemoabsorption.

The Russians and East Germans have used BI extensively in the treatment of various gynecological disorders ranging from adnexitis through endometriosis to disruptions of the menstrual cycle. One Russian study of 215 women treated for these conditions found that UBI had an analgesic, detoxifying, and anti-inflammatory effect.[145]

Frick found that patients he treated for migraine headaches experienced a normalization of menstrual cycle and conception as a side benefit. He reported another study in which 9 out of 30 women desiring children achieved regular menstrual cycles after UBI and 10 more after hormonal treatment following UBI. Frick treated two further cases of anorexia connected with amenorrhea; both conditions resolved following UBI.[146]

A Russian study of the effectiveness of UBI in polycystic ovary syndrome followed 119 patients (54 closely) and 23 controls. Perhaps the most striking effect was that 25 of 29 women complaining of headache improved (the headaches either disappeared or diminished). 29 of 41 with amenorrhea achieved regularization of menstrual cycle, 7 of 24 complaining of infertility became pregnant, and 8 of 42 complaining of hirsutism experienced an improvement. 12 of 30 overweight women lost 3-24 kg in the 3 weeks following UBI therapy without any additional intervention; of them 4 soon regained this weight. Laboratory findings included a disappearance of hyperandrogenism and a tendency toward normalization of secretion of gonadotropins.[147]

---

145 Ibid.

146 Frick (1989), p. 74

147 A.P. Kalinin et al., "UBI in the Treatment of Polycystic Ovary Syndrome [Russian]," Problemy Endokrinologii 38 (1992), No. 6, pp. 19-21

Pregnancy and Birth

While there are few reports from the early Americans regarding the use of UBI in pregnancy, the Russians have treated many problems of mother and fetus during pregnancy with UBI and LBI. In the course of these applications, they have found no evidence that BI has any harmful effects on the fetus. To the contrary, it has consistently proven to be less toxic than any competing drug.

In a major Russian clinical trial, 91 patients with preeclampsia (EPH gestosis)--a disease of apparently immunological origin involving edema, proteinuria, and hypertension--in the third trimester were treated with LBI. The researchers focussed on the effects of LBI on lipid peroxidation of erythrocyte membranes, which was felt to be a key factor in the condition. They treated 61 patients with LBI at 9 milliwatts (N.B., somewhat higher than the normal dose of 1-5 milliwatts) for 20 minutes for 7 days in a row in addition to standard drugs, while 30 control patients received standard drugs only.[148] 11 women with uncomplicated pregnancies served as a second control group.

The researchers found that the addition of LBI helped stabilize erythrocyte membranes, improved microcirculation and blood rheology, significantly decreased hemolysis, increased diuresis, resolved edema more rapidly, dramatically reduced proteinuria (to 0.24 g/l compared to 0.82 g/l ($p<.001$) in the controls), lowered blood cholesterol (thereby lessening the damage to erythrocyte membranes), and more rapidly alleviated hypertension and stabilized arterial pressure compared to the control group's results.

Babies born to the LBI group had better Apgar scale values. 20 percent of the LBI group mothers required caesareans, half of them for severe preeclampsia. 10 percent required placental delivery because of treatment failure, for a total of 20 percent that had less than adequate responses to combination drug therapy plus LBI. Of the control group, 31 percent required caesareans--all of whom had severe preeclampsia. 30 percent of the control group required placental delivery, for a total of 61 percent that had less than adequate responses to combination drug therapy alone. In other words, the inclusion of LBI cut the rate of unsuccessful response to treatment according to this measure from 61 to

---

[148] These included hypotensive, sedative, and spasmolytic drugs and vitamins; in moderate and severe cases they also included infusion therapy, diuretics, anticoagulants, and vasodilating drugs.

20 percent.[149] It permitted the treatment group to have overall outcomes for mother and fetus that came close to those for the normal controls.

It is worth noting that preeclampsia is the second highest cause of maternal mortality in pregnancy in the United States; it retards fetal growth and can cause fetal death. The only treatment for severe cases is placental delivery.

A series of Russian researchers have carefully studied the effects of BI on the fetus.[150] None of them discovered a single harmful effect, including any sign of mutagenicity. In one study, the researchers found that BI stimulated hypothalmo-neurosecretion that restored the central mechanisms of regulation of the defensive-adjustment reaction to hypoxia. It was also found that the inclusion of BI in the combination therapy of newborns with asphyxia shortened their hospitalization time, improved their rate of weight gain, and enabled them to avoid retardation in growth. The use of BI in the treatment of anemia of pregnancy was shown to have favorable effects not only on the mothers but also on the fetuses according to cardiomonitoring.[151]

Studies also showed that the inclusion of BI and abdominal depression in combination therapy in conditions of placental deficiency reduced perinatal mortality in high-risk cases from 72 percent to 6.6 percent. It lessened complications at birth and increased placental blood flow by 10 percent in contrast with controls.[152]

Russian pediatricians confronting the problem of drug-resistance to antibiotics conducted a study of 52 neonates up to 12 hours old in critical condition with suppurative or inflammatory disorders.[153] 35 of them had pneumonia, of whom 17 had aspirational pneumonia. 4 had sepsis. 13 had meningitis and ventriculitis. All received 3 courses of UBI in a row, in a dose equivalent to that for adults, adjusted for their body weights.

---

[149] A.S. Bednarskii et al. "Use of Intravascular LBI in the Combination Therapy of Preeclampsia [Russian]," Akusherstvo i Ginekologiia (1995), No. 6, pp. 18-22

[150] Maltsuyev (1990), p. 8

[151] Curiously, one of the first Russian efforts in BI therapy was undertaken as early as 1941 when a treatment of 35 gynecological patients with septic conditions was adjudged a success. As elsewhere, however, this finding was not followed up on for many years. See E.I. Sidel'kovskaya, "Treatment of Septic Conditions with the Infusion of Blood Irradiated with Ultraviolet Rays [Russian]." Akusherstvo i Ginekologiia (1941), No. 7-8, pp. 14-15.

[152] Maltsuyev (1990), p. 8

[153] N.M. Mezhirova and V.V. Danilova, "UBI for Suppurative-Inflammatory Disorders in Neonates in Critical Condition [Russian]," Anesteziologiia i Reanimatologiia (1993), No. 2, pp. 50-52

The antihypoxic effects of UBI became apparent early on in 28 cases, and 12-16 hours after treatment generally beneficial results became evident (though it is not reported in how many of the babies). The infants became more active and stopped having labored breath. 7 of them showed signs of emerging from coma. Some of the patients were able to go off inhalators. Both umbilical suppuration and conjunctivitis decreased after UBI in almost all infants. The babies' hematological indicators also showed a tendency to normalize that was greater than that of controls'. Unfortunately, the researchers did not include many details (e.g., outcomes) that would permit one to draw firm conclusions from this study.

Regardless of the weaknesses of certain of these studies (not the preeclampsia one!), the general pattern seems clear. Timely use of BI can be of great assistance to both mother and fetus during pregnancy. There is no evidence that BI has any short- or long-term toxic effect on either fetuses or neonates, in contrast to many drugs.

BI's effectiveness against viruses would presumably make it an excellent treatment of congenital CMV infection, a major cause of severe damage to fetuses and damaging sequelae among infants. In the United States, this disease affects 1 percent of all newborns and costs $1.86 billion annually.

Sexually Transmitted Diseases

Females are more at risk from STDs than are males, for many reasons. Their complications include infertility, adverse pregnancy outcomes, chronic abdominal pain, cervical cancer, and severely damaging sequelae such as gonococcal conjunctivitis that results in blindness. The sexual revolution and the spread of HIV and related infectious diseases have made STDs far more prevalent than previously, and new strains and drug resistance have emerged to complicate the picture. In the developing countries in particular, the subordinate role of married women puts them at risk of infection from promiscuous husbands.

BI appears to be effective in a wide range of STDs, including gonorrhea, bacterial vaginosis, candidiasis, and others. There are no reliable reports on its effectiveness in herpes simplex genitalis, condyloma acuminata (human papilloma virus-associated genital warts), or chlamydia trachomatis. Nonetheless, anecdotal reports and clinical studies of its effect in related indications--e.g., on herpes eye infections and on plantar warts (HPV)-- suggest that it might be effective against the first two.

The early American UBI practitioners treated both acute and chronic gonorrhea with success. They also found UBI effective against gonorrheal arthritis. But the advent of antibiotics led even some of them to set aside UBI for these indications, presumably because it was easier to dispense pills.

Miley used UBI in two cases of syphilis with chancre inguinal lymphoderropathy once weekly for 4-6 weeks.[154] When he noted no effect, he ceased using UBI. In light of what is now known about the toxic profile of BI, however, this might have been an example of a case in which even an experienced UBI practitioner did not use a sufficiently high dose, nor did he treat at the optimal intervals. There is evidence from a Russian study of BI in rheumatoid arthritis that daily treatment is more effective. In regard to dose, a different approach for such an serious, hard-to-treat indication as syphilis would have been to treat with UBI at intervals of several days while observing the patient carefully until the point at which the first symptoms of toxicity occurred, then to stop if the treatment was having no effect. Had this procedure been followed in these two cases as well as in countless others, many apparent failures of BI might have turned into successes.

Miley did report, however, marked success with two cases of syphilitic optic uritis after UBI treatment. This in turn reinforces the above point. In general, many of the early American doctors appear to have been reluctant to use UBI in an aggressive way for fear that it might cause some hidden damage to their patients. They may have also feared that this would then be used against them by critics and hurt the cause of what they viewed as a wonderful therapy. Two noteworthy exceptions were Rebbeck and Olney.

---

154 Miley (1997), p. 116. It is not clear whether the patients were female.

Pelvic Inflammatory Disease

Pelvic Inflammatory Disease (PID) is any acute or chronic infectious or inflammatory/autoimmune process that affects the cervix (cervicitis), uterus (endometritis), fallopian tubes (salpingitis), and ovaries (oophoritis). It is a major outcome of many infections as well as of the insertion of a catheter, abortion, and surgery. PID can prove hard to treat with drugs and can cause infertility.

Between June 1944 and mid-1947, R.C. Olney treated 631 PID patients with UBI.[155] The 220 he classified as "very severe" had chronic non-gonorrheal PID with massive swelling, with recurrent attacks for from several months to many years. Many had been operated on. UBI monotherapy relieved all symptoms and returned organs to normal condition in 174 (79 percent). 24 patients (11 percent) were "improved," while 22 (10 percent) required operations for ovarian cysts, fibroids and/or abscesses.

Of 238 classified as "moderately severe", 191 patients (80 percent) had complete recoveries following UBI. 24 (10 percent) were markedly improved but continued to have some symptoms. 23 (10 percent) did not continue treatment and the result was not known.

Of 174 patients with "mild" infection with severe dysmenorrhea, 151 (87 percent) entirely recovered after UBI. 17 patients had improved conditions, and 6 did not return for treatment.

17 patients unable to conceive became pregnant after UBI. In 5 cases of threatened abortion, the symptoms subsided promptly after UBI, with the patients continuing to full term and normal delivery.

Olney commented that

> At the present time we do not operate upon any patient with a pelvic disorder unless there is an abscess, tumor, cyst or other organic condition which is not purely inflammatory.

He attributed the advantage UBI had over antibiotics to its ability not only to destroy invading microorganisms but also to stimulate host resistance.

---

[155] R.C. Olney, "Ultraviolet Blood Irradiation Treatment of Pelvic Cellulitis. Knott Method," American Journal of Surgery 74 (1947), No. 4, pp. 440-43

In a Russian study of 23 patients with various kinds of PID and related conditions compared with a group of 24 controls with the same disorders, 1-6 daily treatments with UBI were administered in addition to the drug and physical therapy standard for both groups.[156] Two incidents occurred during the UBI treatments. One patient collapsed at the time of blood-taking, which the researchers ascribed to the rapidity of withdrawal of blood and which was treated successfully with drugs. A second became dizzy and nauseous at the time of the return of blood to her vein, apparently because she could not stand the sight of her own blood. She recovered by herself in 20 minutes. Neither patient had any reactions during successive treatments.

These incidents underline the fact that BI's needles and bloodtaking carry with them a minor risk of such incidents. One of the weaknesses of the originally Knott Hemo-Irradiator and the Russian Isolda UBI device is that they are connected to the patient's arm, so that it is not possible to perform the procedure out of the patient's sight. Of course, the patient can always close his/her eyes.

Pain disappeared, temperature normalized, and pathological signs on examination disappeared much sooner in the standard treatment plus UBI group than in the control group. The UBI group healed in an average of 12 days, while the controls took 21 days.

Another Russian study compared three groups of 30 women each. Group 1 consisted of healthy women. Group 2 had PID and received LBI (7-10 sessions of 25-30 minutes of 5 milliwatts at 632.8 nm) in addition to standard therapy. Group 3 had PID and received standard therapy only. Groups 2 and 3 both had symptoms of pelvic peritonitis. The researchers were particularly interested in the effects of LBI on the hypercoagulatory conditions, including suppression of thrombolysis, that develop in PID.

The inflammation in the Group 2 patients was brought under control in 6-7 days; that in Group 3 was controlled in 9-10 days. Normalization of coagulation and thrombolysis occurred at the same times. There were no coagulatory problems in Group 2 whereas 3 cases of thrombophlebitis of the cubital veins occurred in Group 3.[157]

---

[156] O.A. Mashkin et al., "The UBI Method in the Combination Therapy of Patients with Inflammatory Conditions of the Genitals [Russian]," Akusherstvo i Ginekologiia (1990), No. 10, pp. 58-60

[157] Brill' (1994), pp. 158-63

Conclusion

Many antibiotic treatments exist for STDs and for PID. BI appears to be as good as or better than any of them, but it is unrealistic to think that it will enter into common use in those indications for which easily taken, effective antibiotics are available. It does appear capable of treating several STDs, however, which are not very susceptible to drugs--in particular, HPV-associated genital warts (condyloma acuminata). In poor countries where the latest antiviral drugs are not widely available, BI may prove a suitable substitute. It can also treat cases of drug resistant and recurring infections.

In the treatment of problem pregnancies, BI is an excellent therapy for preeclampsia--far better than any approach currently used in the United States. BI's potential to treat congenital (antepartum) CMV successfully is also of considerable interest. In general, the unusually low toxicity of BI holds out great promise for treating both mothers and fetuses with far more sensitivity and effectiveness than is the case with current drug therapies. And if BI can be so low in toxicity in the treatment of fetuses and newborns, then its use on various pediatric indications would appear to be equally safe.

Chapter 9: Other Uses of Blood Irradiation Therapy

It is quite likely that somewhere, sometime a BI practitioner has attempted to use this therapy on practically every disease, disorder, condition, or other physical or mental situation imaginable. Without any claim to comprehensiveness, here are various additional uses to which BI has been put, from the very major to the near-trivial.

Autoimmune Diseases

With the advance of knowledge about the immune system, autoimmune processes have been implicated in more and more diseases and conditions. Thus Disease-Modifying Anti-Rheumatic drugs (DMARDs) such as methotrexate and azathioprine have joined or replaced glucocorticoids and other treatments as long-term therapies. In contrast to analgesics like aspirin and non-steroid anti-inflammatory drugs (NSAIDs), DMARDs can suppress autoimmune activity long enough for the body to heal itself via normal turnover of cells or at least to limit the damage from the disease.

BI's ability to modulate or destroy the activated T-4 cells that drive such syndromes clearly puts it in the DMARD category. The early American UBI practitioners found that the results were variable in rheumatoid arthritis and many other autoimmune syndromes. Generally, UBI was highly effective in early-stage cases but only erratically effective in advanced ones. There was, however, one important exception: UBI was invariably effective in the difficult indication of ankylosing spondylitis, regardless of the stage of disease.[158]

Rather than survey all the various autoimmune diseases that American, German, and Russian BI practitioners have treated, it seems advisable to focus on two major Russian clinical trials--one of LBI in rheumatoid arthritis, the other of UBI in multiple sclerosis.

The rheumatoid arthritis trial is of special interest both for the curious findings and for the sophistication of the trial itself.[159] 150 patients with RA diagnosed according to American guidelines were randomized into 5 groups:

- Group 1 (28 patients) was treated 2-3 times per week for a total of 4-6 treatments, with the laser at 1 milliwatt.
- Group 2 (25 patients) was treated in the same way except at 6 milliwatts.
- Group 3 (28 patients) received daily LBI treatment for 6 days at 1 milliwatt.
- Group 4 (51 patients) received daily LBI for 6 days at 6 milliwatts.
- Group 5 (18 patients) served as a placebo control. They received "invasive placebo" therapy whereby 2-3 times per week a waveguide was inserted in their veins but the laser was not turned on.

The average age was 47.5. The average length of disease was 9.1 years. 127 patients (85 percent) were women; 23 (15 percent) were men. 27 were in stage I of RA, 50 in stage II, 55 in stage III, and 16 in stage IV. All patients received 1-2 NSAIDs before and during BI therapy; none had received DMARDs for 6 months prior to the trial. Several days after the end of therapy, the results were assessed according to seven criteria, which served to form a therapeutic index (TI).

The results were tabulated thus:

---

[158] Miley (1997), p. 192

[159] Zvereva et al. (1994), pp. 29-32

Table 9.1. Results of Treatment of RA with LBI

Result	1	2	3	4	5
significant improvement	12/43	6/24	16/57	25/49	2/11
TI	0.606	0.537	0.589	0.599	0.515
improvement	8/28.5	11/44	8/28	19/37	12/67
TI	0.261	0.318	0.285	0.332	0.295
no change	7/25	6/24	1/3.5	6/12	4/22
TI	0.101	-0.062	0.151	0.040	0.130
deterioration	1/3.5	2/8	3/11	1/2	0/0
TI	-0.659	-1.114	-0.293	-1.011	-
average TI	0.336	0.165	0.401	0.387	0.283

Several important results emerged:

- with $p<0.05$, Groups 3 and 4, which had received daily LBI, significantly outperformed Groups 1 and 2. The impact of daily treatment appeared to be greater than the impact of the level of irradiation;
- although there were no side effects of LBI, in all groups in which it was used there were cases of patients who deteriorated during its use, though not necessarily in response to LBI;
- in Groups 3 and 4, women performed better than men, and patients with early-stage disease performed better than others. 6 patients who did particularly poorly had an average disease duration of 13.9 years, while 8 who did particularly well had a duration of 3.58 years; and
- LBI appears capable of provoking flareups in patients with progressive involvement of more and more body components in the disease process and in patients with progressive destruction of cartilage.

The researchers recommended that LBI be administered in daily doses of 1 milliwatt.

What else can be said about this trial? First, it is clear that a careful study of how effective the randomization process had been in creating equivalent groups would be advisable, given the strong dependence on early stage of disease for effectiveness. Second, the tabulated results suggest that the 6 milliwatts in Group 4 may well have been marginally more effective than the 1 milliwatt in Group 3. Third, BI clearly can have a strong placebo effect: 2 of 11 patients in Group 5 scored a significant improvement. Fourth, the pharmacological effect of BI is also important: overall, Groups 3 and 4 decisively outperformed Group 5.

One explanation for the superiority of daily doses of LBI over doses at greater intervals is that, before the stimulation of the blood cells dies down, a new dose is added, thus causing a higher degree of activation. Another explanation is that the body responds more favorably to daily irradiation because it resembles the daily irradiation of the sun, to which body cells are attuned.

A cautionary note: RA is a very tricky disease with multiple manifestations that can be hard to explain. It can also have a significant psychological component. Thus too much weight should not be accorded to any single study, even to such an exemplary one as this.

A second Russian study treated 172 patients with autoimmune conditions with LBI. 132 had RA, 32 had systemic lupus, and 17 had systemic scleroderma. A control group included 43 patients with RA who received standard drug therapy. 52 of the LBI patients also received an inhibitor of lipid peroxidation. All LBI patients also received non-steroid anti-inflammatory drugs. Reduction in the clinical and laboratory indicators of inflammation occurred more rapidly in the LBI group, as did the level of circulating immune complexes. After 3-4 sessions of LBI, a temporary exacerbation of "joint syndrome" and general weakness accompanied by subfebrile temperature occurred. Those patients receiving antioxidant therapy, however, did not experience these effects.[160]

---

[160] Izhevsk (1994), pp. 139-40

BI therapy has been often tried on patients with MS, but the results have generally been disappointing. In theory, BI's anti-inflammatory property whereby it suppresses excessive T-cell activation should be very effective in suppressing the demyelination in MS, at least in early-stage cases. This appears to be the mechanism of action of drugs like methotrexate which have been shown to have effect in slowing the progression of MS. BI's general failure against MS suggests that its secondary radiation might be less effective in penetrating the region where the demyelination takes place.

However, of 5 patients treated by Barger, one improved markedly and another made a dramatic recovery from a terminal stage of MS after 4 sessions of UBI and lived a relatively normal life for some years.[161]

In a Russian clinical trial, physicians irradiated the cerebrospinal fluid of 26 MS patients with UBI twice, with a one-week interval.[162] They obtained good results with 14 patients and no results with 12. The responding patients tended to have less aggressive cases of MS at the outset.

However, this approach did not take advantage of the whole-body physiological effect of BI. Rather, the researchers assumed that UV-induced changes in the membranes of blood cells were the central mechanism of BI, and they reported an increase in the activation of the immunological components of the cerebrospinal fluid in the responders but not in the non-responders, which would appear paradoxical in light of the known immunological basis of MS. In addition, they only irradiated 5-7 ml of the fluid.

This clinical trial, therefore, is of some limited interest for the comparative treatment of MS, but it does not shed much light on the strengths and weaknesses of standard BI in the treatment of this disease.

---

[161] Miley (1997), p. 201

[162] N.A. Totolyan et al. "UV Irradiation of Cerebrospinal Fluid in the Treatment of Patients with Multiple Sclerosis [Russian]," Zhurnal Nevropatologii i Psikhiatrii Imeni S.S. Korsakova (1993), No. 4, pp. 9-14

Parkinson's Disease

No successes have been reported in the treatment of Parkinson's with BI.

Migraine Headaches

BI has the advantage over certain drugs that it appears to pass the blood-brain barrier with little difficulty, though how it does so is not clear. It has the advantage over local irradiation of the brain and other local treatments of approaching the scene via the more natural mode of the blood.

Many BI practitioners have employed it against migraine headaches. Miley treated 12 patients with longstanding migraine and found "striking improvement" in 11 from the outset, though he had to treat 7 with maintenance therapy every 6-10 weeks for up to 3 years, during which there was no recurrence. He also used UBI in headaches following whiplash trauma (with success) and in muscle-contraction and torticollis-associated headaches (with only occasional success).[163]

The East German Taubert used UBI extensively in the treatment of migraine headaches.[164] He found it of considerable benefit and performed several comparative studies with placebo to determine whether the effect was psychological (very possible with headache) or physiological. He concluded that a placebo effect was often present, but his results show "UBI is clearly superior to placebo treatment and thus its efficacy is based on physiological effects."[165]

In a dissertation Taubert cites, 21 migraine patients received UBI in a double-blinded study. Of them 2 were subsequently free of complaints; 5 noted significant improvement; 6 experienced some improvement, and 8 had no change.[166] In general, his studies reveal that

---

[163] Miley (1997), pp. 197-98

[164] K. Taubert, "Ultraviolettbestrahlung des Blutes (UVB) bei Migraene," Zeitschrift fuer aertzliche Fortbildung (Jena) 85 (1991), No. 1-2, pp. 43-8

[165] Ibid., p. 46

[166] M. Ziepert and M. Ziepert, "Zur Wirkung der Ultraviolettbestrahlung des Eigenblutes bei der Migraene," Friedrich-Schiller-Universitaet Jena, Medizinische Fakultaet. Dissertation A, 1985, cited in Taubert (1991), p. 45

some 60-80 percent of migraine patients benefited from UBI, some of them even becoming free of headaches entirely.

How exactly BI might work in various headaches is not clear. Its effects on freeing up microcirculation, on vasodilation, and on relaxation of the autonomic nervous system could all play a role.

Non-Healing Wounds

In keeping with BI's pattern of action against stomach ulcers, it has frequently been found effective in the treatment of skin ulcers and is effective in healing burns. Miley published one report of UBI in the treatment of 6 cases of non-healing wounds, all of which appeared to respond very well to UBI.[167]

In one of these cases a 39-year old woman was diagnosed with chronic staphylococcus cellulitis, chronic osteomyelitis, chronic arthritis, chronic cholecystitis, and contraction flexures of both arms and both legs. During several years of treatment, she had draining lesions on both arms and legs, an abscess of the cervix, and other complications. At times she appeared "apparently hopelessly incapacitated." At irregular intervals 18 UBI treatments were administered. After receiving them, she noted that she felt stronger for several weeks. Gradually her condition improved. Her lesions healed, as did her other conditions. During the last year she was under observation, she was free of all "clinical signs" of chronic cellulitis, chronic arthritis, chronic osteomyelitis, and chronic cholecystitis. Her sedimentation rate and laboratory findings were normal. Only a flexure contraction of the left elbow remained. She was able to resume a "limited normal life."

Dermatology

BI has been used in dozens of skin disorders. It appears to be most effective in the treatment of furuncles, carbuncles, and other

---

[167] George Miley, "Ultraviolet Blood Irradiation Therapy (Knott Technic) in Non-Healing Wounds," American Journal of Surgery 65 (1944), No. 3, pp. 368-372

abscesses. In 1950 Hermann Fruehauf reported on the experience that he and other German physicians had had with BI. Their technique was to irradiate some 10 cm of the patient's blood with UV, then inject it locally. He claimed that his treatment of furuncles and carbuncles was invariably successful, but that it was necessary to give at least 10 intramuscular injections. For certain indications, he gave as many as 30-50, usually on a twice-weekly schedule.[168]

A Russian trial of scleroderma patients with Raynaud's syndrome compared 28 patients receiving LBI with 30 controls who received standard drug therapy. 43 percent of the first group experienced significant clinical improvement; 50 percent benefited somewhat; and 7 percent did not respond. Among the controls, 33 percent had significant improvement; 16.7 percent benefited somewhat; and 50 percent did not benefit. One patient's condition worsened.[169]

In a trial that compared two groups of patients after operations for carbuncles and phlegmons, Russian researchers studied a number of indicators of nonspecific resistance. Group 1 included 16 patients who received 5-12 UBI treatments in the weeks following operations. Group 2 included 14 who did not receive UBI following operations. Group 3 consisted of 25 healthy controls.

The patients in Group 1 had reduced pain, suppuration, and hyperemia of the surrounding skin as well as normalization of body temperature much sooner than did those in Group 2. On two indicators of nonspecific resistance, the Group 1 patients equalled or surpassed the healthy controls by the end of their treatment whereas the Group 2 patients lagged behind. The C-reactive protein level of the Group 1 patients with carbuncles sank from 3.1 to 0.2 by the end, while that in Group 2 patients dropped from 2.8 to 1.4 only, i.e., these patients still showed mild signs of inflammatory activity. In the patients with phlegmons, the C-reactive protein level sank from 2.9 to 0.8 in Group 1 and from 3.2 to 2.0 in Group 2.

The researchers commented that they had treated 726 patients with various diseases for a total of 3,500 sessions of UBI. A significant healing effect was found in 84 percent of the cases, and there was no case of complications.[170]

---

[168] Hermann Fruehauf, "Therapie mit ultraviolettbestrahltem Eigenblut," Medizinische Klinik 45 (1950), No. 46, pp. 1469-72

[169] Izhevsk (1994), p. 60

[170] Ganelina and Samoilova (1986), pp. 25-30

In a clinical trial of UBI on children (5-14 years old) with various skin conditions, the researchers deliberately introduced a filter that reduced the dose to 1/10 of normal.[171] They divided the patients into two groups. Group 1 included 11 patients with chronic pyodermatitis or progressive psoriasis. Group 2 included 17 patients with allergodermatitis in whom the basic syndrome was recidival pyodermatitis. Group 3 consisted of 10 healthy controls. Groups 1 and 2 received 5-10 sessions of UBI.

In Group 1, the patients with pyodermatitis all attained full remission after 5 sessions. Of those with psoriasis, 2 had full remissions, 2 had significant improvements, and 2 had stabilization of their disease.

In Group 2, the patients at first experienced general improvement. But then their erythema flared up again, fresh pustules appeared, and the effect worsened with more treatment--though 6 were reported as attaining complete remission. This remission did not last 3 months, however.

In contrast to other therapies, UBI had the advantage in both groups (presumably aside from the 11 worsened cases in Group 2, though this is not clear) of normalizing the functional activity of the vegetative nervous system and raising the level of the antibodies against staphylococcus. The researchers noted that Group 1 had experienced full remissions in many cases with only 1/10 of the normal dose of UBI. However, on the analogy of BI's action in rheumatoid arthritis and other disorders, it seems possible that the very low dose was to blame for the poor outcomes of Group 2. In effect, the researchers had managed to give the children just enough UBI to provoke the inflammatory process but not enough to suppress it.

When Dr. Knott and his medical collaborators established the standard dosage of the Knott Technique, they were basing their recommendations on extensive experience. They evidently knew what they were talking about.

Other skin disorders treatable with BI include eczema, allergic skin reactions, keloid, vitiligo, psoriasis (one authority considers PUVA superior to BI for this indication, however[172]), acne, erysipelas, pemphigus, dermatitis herpetiformis, and chronic urticaria.

---

[171] Ibid., pp. 123-32

[172] Frick (1989), p. 71

Diabetes mellitus

Given the energy-based nature of BI and its consequent action parallel to glucose metabolism, it would be expected that BI should have some influence over diabetes. Several of the early American UBI practitioners used it on diabetic patients. Miley found that he was able to reduce the blood sugar of one diabetic from 350-400 mg to 140-150 mg with two sessions of UBI, but that if 3-4 weeks passed without UBI, the level rose to 350-400 mg again. He maintained an otherwise insulin-dependent patient on a UBI regime without any insulin for 18 months. He did recommend, however, that UBI not be given without reducing insulin because of the danger of hypoglycemic shock.[173]

Frick found that UBI was somewhat helpful in lowering glucose level and in permitting better utilization of insulin. He also obtained good results in the treatment of 58 patients with diabetic retinopathy with UBI. Erythrocyte aggregation was suppressed. About 70 percent of patients in stage II experienced an improvement of vision, and 20 percent in stage III.[174]

Another East German study of the impact of UBI on blood glucose levels in 18 patients with vascular disease (14), MS (1), RA (2), and thrombophlebitis (1) found a uniformly positive effect in reducing glucose levels. Measured 30 minutes after UBI treatment, the average level was 28 percent lower. This result included 10 patients with diabetes. One week later, the researchers "treated" the patients again, but this time in a double-blinded manner with the UBI device deliberately not turned on. The average drop in blood glucose level 30 minutes after this placebo treatment was 6.8 percent.[175]

A Russian team reported good results with UBI in the treatment of non-insulin dependent diabetes but lack of success in treating insulin-dependent patients. But they only used a single dose of UBI in a clinical trial of the impact of UBI on lipid peroxidation of erythrocyte membranes, so that these conclusions lack weight.[176]

---

[173] Miley (1997), pp. 145-6

[174] Frick (1989), p. 66

[175] F.-W. Gaensicke and S. Wiesner, "Der Einfluss der Ultraviolettbestrahlung des Eigenblutes auf den Glukoseumsatz," Zeitschrift der Physiotherapie 27 (1975), pp. 435-6

[176] A.L. Rakov et al., "UBI in the Combination Therapy of Diabetes Patients [Russian]," Klinicheskaia Meditsina 60 (1991), No. 8, pp. 95-99

Other Russian researchers conducted a trial of 63 patients with progressive microangiopathy, macroangiopathy, and other complications of diabetes. The patients received 7-10 doses of LBI at 1.5-2.0 milliwatts for 30 minutes in addition to standard drugs. The combination therapy reduced parasthenia and pain in extremities, polyuria, and other problems. The glycemic profile of 82 percent stabilized, and maximal deviations were reduced by 15-20 percent. After 3 months, the level of glycosylated hemoglobin was found to be lower. The researchers considered that LBI had favorably influenced the overall well-being of the patients, but the lack of controls makes it difficult to assess this finding.[177]

The other area in which many successes have been scored with BI treatment of diabetes is in reducing damage from diabetes-induced obstruction of the lower extremities.

While the advisability of employing BI to lower high glucose levels in diabetes remains an open question, therefore, BI certainly can play a highly beneficial role in treating ocular and vascular complications of this disorder.

Anemia

BI has been used by various practitioners on anemia with varying results, but it is difficult to generalize because there are so many different kinds of anemia. A Russian study from 1937, the first Russian report on UBI, found that in the treatment of 86 cases there was a minor advantage to UBI. But the results were no greater than those from simple autotransfusion.[178]

---

177 Izhevsk (1994), pp. 122-3

178 A.N. Filatov and G. Kasumov, "Experimental and Clinical Observations on the Question of the Transfusion of Blood Subjected to Irradiation with Ultraviolet Rays in Anemias [Russian]," Vestnik Khirurgii Imeni I.I. Grekova 49 (1937), pp. 3-10

BI conveys the effects typical of an oxygenation therapy--most notably in its action in vascular disorders, but in many other ways as well. In a 1939 study of 83 treatments with UBI, readings of blood oxygen rose from 7.0 to 11.1 when measured immediately before and 10 minutes after treatment--an average of 58.6 percent. In a second group of 14 UBI treatments, the second measurement was made 30 minutes after UBI. In this case, the average combined oxygen value of 9.7 percent dropped to 8.8 percent after irradiation. In a third group, one month subsequent to 27 UBI treatments, oxygen values had risen from 5.8 percent to 8.7 percent, an average increase of 50 percent.

The explanation of the drop of the second group may have been that UBI has its most significant oxygenation effect in cases where the initial value is low; or it might be explained as a result of a drop to a point not very far removed from the initial point after a short surge in the value. The one-month results then could be interpreted as the results of the overall normalization of the blood arising from the overall healing process following therapy rather than to any special persistence of high oxygen levels.[179]

A German study found a significant difference between oxygenation therapy with and without UBI to be the marked rise in hydrogen peroxide following UV irradiation, whereas mere oxygenation did not affect hydrogen peroxide levels.[180]

In a 1970 study, American physicians determined that venous blood oxygen in 21 patients with hypoxia rose by a mean of 19.8 percent following UBI as measured 24 hours later. When the same patients received a second UBI treatment, the blood oxygen rose by another 12 percent on average, measured 24 hours after treatment. Thus the overall rise was 25.6 percent above the initial pre-UBI value.[181]

---

[179] Miley (1939), p. 873

[180] F. Wennig, "Blutveraenderungen durch UV-Bestrahlung in O_2-Atmosphaere," Wiener Medizinische Wochenschrift 106 (1955), No. 49, pp. 1015-8

[181] Robert C. Olney and Armand C. Grez, "Treatment of Hypoxia: A Preliminary Report on 21 Cases," Abdominal Surgery 12 (1970), No. 3, pp. 45-6

Scores of practitioners have sought to treat cancer with UBI and LBI. If BI were powerfully and consistently effective in destroying tumors, we can be sure that we would have heard about it. In fact, the consensus is that BI is not effective against cancer. This appears to hold even for hematological malignancies, which in theory should prove susceptible to a bloodborne therapy like BI.

However, the last word has not been spoken. According to a few isolated reports, in certain circumstances BI *can* be effective against cancer. Of course, the same has been claimed about practically every serious drug ever invented. Still, the formidable action that BI has in many difficult indications suggests that it would be wise to take a closer look.

What is the evidence that BI might be effective?

Dr. Olney wrote a pamphlet describing 5 cases in which he had successfully treated cancer with UBI. As in all such accounts, it is impossible to tell whether UBI worked or whether misdiagnosis, prior therapy, or placebo effect was the cause of the seeming disappearance of tumors. So, too, Dr. Douglass reports a case in which UBI was apparently successful. Both Olney and Douglass ascribed their successes to the oxygenation effect of BI.[182]

Two other possible mechanisms might be: the concentrated effects on tumors of the secondary radiation emitted by the irradiated cells and the overall stimulus BI gives to the body's immune defenses.

It is likely that, if BI has any consistent effect on tumors, it is in the very early stages of their growth when they are most vulnerable. Thus BI may be effective as prophylaxis against breast and ovarian cancer. Still, it should not be ruled out that in the right circumstances, and with a large enough dose, BI might have direct action against a more advanced tumor--or rather, BI might sufficiently stimulate the immune system that it rises up and destroys the tumor.

In Germany, one physician used Hemo-oxidative Therapy (HOT: UBI plus oxygenation) as standard therapy on 8,000 cancer patients.[183]

Aside from destroying tumors, there is a good argument for using BI as an adjuvant therapy in cancer. It can reduce the side-effects of

---

182 Douglass (1996), pp. 139-50

183 Frick (1989), p. 75. See Doerfler, J. "Haematogene Oxydationstherapie zur Krebsbehandlung," Krebsgeschehen 6 (1972), pp. 136-43, cited in Frick.

chemotherapy and in theory suppress such accompanying syndromes as Idiopathic Thrombocytopenic Purpura (ITP).

In a Ukrainian trial of patients with stomach and duodenal cancer, LBI was administered 7-10 times to 76 patients in the pre- and early operative periods. The post-operational development of infections in the control group was 19.7 percent; in the group treated with LBI, it was 4.9 percent. LBI also made possible a significant reduction in medications.[184]

In another Ukrainian study, the researchers noted that Soviet oncologists had been very reluctant to use BI for fear that UV irradiation might stimulate the growth of normal and tumor cells; but they pointed to several studies that showed this fear to be groundless.[185]

In their study, 367 patients who had undergone radical thoracoabdominal operations for cancer of the esophagus and stomach were included. 207 received 2-4 treatments with UBI as part of combination therapy following their operations. They noted that 13 patients said that they had a metallic taste in their mouths and felt sleepy during UBI. 160 patients receiving standard therapy constituted the control group.

In the UBI group, the incidence of arterial blood pressure changes and arrhythmia dropped by 25 percent. Suppurative-inflammatory complications occurred in 10 percent of the UBI group and 13.1 percent of the controls. The level of peptides of average molecular mass was significantly lower in the UBI group, and there was a concomitant reduction in the severity of pancreatitis following the operations.[186]

In addition, BI's antiviral capability has the potential to reduce the rate of common cancers by destroying viral infections before they can have a long-term oncogenic effect on body cells. BI's action versus human papilloma virus (HPV) in genital warts, for instance, can reduce the incidence of cervical cancer, while its antiviral effects in chronic hepatitis B and C can suppress the virus before it leads to liver cancer. In a telling incident, Douglass reports treating with UBI an elderly woman for cancer only to discover that, while the treatment had no

---

[184] N.A. Yal'chenko and V.D. Lagutin, "The Influence of IV LBI on the Activation of the Sympatho-adrenal, Antioxidative, and Immune Systems in Patients with Stomach and Duodenal Cancer [Russian]," Klinichna Khirurhiia (1996), No. 1, pp. 36-39

[185] Among them a doctoral dissertation at Leningrad University: I.I. Firulina, "Changes in the Growth-Stimulating Properties of Human Blood Proteins after UV Irradiation in Therapeutic Doses [Russian]," (1987).

[186] N.N. Zhurko et al., "The Use of UBI after Operations in Patients with Tumors [Russian]," Klinicheskaia Khirurgiia (1992), No. 6, pp. 12-15

effect on the cancer, it caused a plantar wart of 25 years' standing to disappear.[187]

BI is a virus killer, not a tumor killer.

It is highly unlikely that BI might be an amazing cure for cancer. Rather, a proper assessment of BI's potential in various areas of oncology and hematology would permit it to be carefully employed in the areas of cancer medicine where it might do the most good.

Burns

In a study of 87 patients with burns covering 3-60 percent of body surface, of which 56 were of IIIB-IV degree from 2-38 percent of body surface, Russian researchers reported that soon after the first procedure, the patients felt euphoric. In the course of the first few days their mood improved, sleep normalized, their appetites rose, and the intenseness of pain diminished, which enabled many of them to go off narcotic analgesics. Severe pneumonias rapidly disappeared. Epithelization of the surface of II and IIIA degree burns took place, which shortened patients' hospitalization from 33.9 days (without UBI) to 26.2 days (with UBI). Improvements in immune status permitted those patients with IIIB and IV degree burns of up to 10 percent of body surface to leave the hospital significantly sooner than under standard treatment.

No details were provided regarding the long-term impact of UBI treatment on more extensively burned patients.[188]

Biological Warfare Agents

BI's ability to destroy microorganisms and deactivate toxins makes it a promising treatment for virtually every currently stocked agent. This includes anthrax, Brucellosis, Yersinia pestis, Q Fever (rickettsia-like), smallpox (variola virus), viral encephalitides, viral

---

[187] Douglass (1996), p. 150

[188] Ganelina and Samoilova (1986), pp. 39-49

hemorrhagic fevers, botulinum toxins, and Staphylococcal Enterotoxin B (SEB).[189]

BI in Anesthesiology

In Russia BI has been found to be very helpful in overcoming operational and post-operation stress.[190] By relaxing the autonomic nervous system, it reduces vasospasm, overcomes the imbalance of water and electrolytes, and has a pronounced analgesic effect. There is also a rationale for using BI in cases of pain caused by cancer and other conditions. BI is currently being used by a reputable American anesthesiologist for this purpose.

BI in Ophthalmology

BI has shown itself to be highly effective against viral and autoimmune eye disorders. An American study that included both was discussed in Chapter 4. Several more recent studies are also worth mentioning.

According to an East German study, 73 percent of patients with iridocyclitis and uveitis were cured in 4-5 weeks, and 15 percent more after further treatment with UBI. In some cases, an exacerbation would occur after 1 or 2 treatments, but persistent UBI resolved this and brought the best results of all. The practitioner could continue

---

189 This list of current agents is drawn from David R. Franz et al., "Clinical Recognition and Management of Patients Exposed to Biological Warfare Agents," JAMA, Aug. 6, 1997, Vol. 278, No. 5, pp. 399-411. UBI was tested during World War II as an oxygenation treatment for divers' bends, but it is not clear whether BI has ever been studied by the U.S. Department of Defense or other militaries as a treatment for biological warfare agents, though it is in regular use by Russian military hospitals for various diseases, burns, etc.

190 M. Y. Avrutskii et al. "The Influence of Low-Intensity Laser Irradiation on the Fundamental Biological Processes and Homeostasis in Sick People [Russian]," Anesteziologiia i Reanimatologiia (1991), No. 5, pp. 74-79

antibiotics and glucocorticoids during UBI, but removing them entirely brought more satisfactory results.[191]

A Russian team achieved good results using LBI at 632.8 nm on 16 patients (20 eyes) with relapsing chronic keratitis and uveitis in the presence of immunological problems. All 16 patients responded with a sharpening of visual acuity, resorption of corneal precipitate, and other improvements. However, there were no controls.[192]

An Astrakhan opthalmologist reported on his dual approach of using UBI plus subconjunctival injection of UV-irradiated blood in the treatment of keratitis of the cornea and conjunctiva. 38 patients in Group 1 received UBI, 24 in Group 2 received UBI and UV-irradiated injections, 16 patients received standard drug therapy and injection of non-irradiated blood, and 19 patients in Group 4 received standard drug therapy and autotransfusion therapy. The diagnoses included viral, traumatic, and bacterial keratitis as well as a range of other related disorders. All were severe or moderately severe cases.

The patients in Group 2 (UBI plus subconjunctival injections of UV-irradiated blood) had superior outcomes in terms of sharpness of vision after therapy and fewer days of hospitalization. The other three groups were roughly equal in terms of sharpness of vision, but Group 1 (UBI) was better than the other two in terms of length of hospitalization.[193]

One problem with studies like this is that they don't adequately address the question of comparative effectiveness because one group (Group 2: UBI plus subconjunctival injection of UV-irradiated blood) received more therapy than another (Group 1: UBI). In effect, Group 2 received a higher dose, and therefore it is not surprising that it did better. But that does not answer the question of the relative merits of the two approaches. What is needed, rather, is a formula--however rough--for determining equivalent amounts of UBI and local UV-irradiated blood, then to compare the results of these two approaches. This methodological weakness turns up repeatedly in the Russian literature and, to some extent, in the American and German studies of BI as well. Many physicians feel more comfortable adding new elements to an

---

[191] Frick (1989), p. 69

[192] Y. V. Yelisereva et al., "Intravascular LBI in the Treatment of Various Ocular Disorders [Russian]," Vestnik Oftalmologii 110 (1994), No. 2, pp. 23-24

[193] L.D. Zhivotovskii, "The Use of UV-Irradiated Blood in the Treatment of Keratitis [Russian]," Oftalmologicheskii Zhurnal (1990), No. 7, pp. 420-3

existing combination therapy, though, so it is difficult to gain a clear picture.

In theory, given truly equivalent doses, BI can obtain results superior to local therapy because the blood cells emitting the secondary radiation enter the eye via the normal, natural route and in exactly the right dose. Once the inflammatory problem is solved, they disappear. However, it should also be possible to provide a highly effective treatment with subconjunctival injections. One question would be: what danger might such an approach contain of local toxicity because of overdose? Systemic UBI is free of this danger; it is an open question whether local injection of irradiated blood is.

Lastly, ophthalmologists in Voronezh reported on the treatment of 59 patients with primary glaucoma with LBI at 680 nm and 0.5 milliwatt power. In comparison with unspecified controls who received standard drug therapy, LBI was superior in regard to improvement of visual acuity, reduction of intraocular pressure, and other parameters.[194]

Neurology

A Russian study of 65 patients with various neurological and physical disorders diagnosed as originating in stress found that 47 experienced a significant effect from 6 treatments with LBI, 12 had a minor effect, and 6 had no response. In contrast to the emphasis on the improvement in cerebral circulation found in other studies of the influence of BI on neurological disorders, this study emphasized the normalizing influence of LBI on the sympathetic and parasympathetic nervous systems. LBI lessened anxiety and increased intellectual activity, but it had less effect on patients who had developed advanced symptoms.[195]

In general, BI has been used by American, German, Russian, and Ukrainian physicians on a wide range of neurological disorders. Some of the studies were reported in Chapter 6. While the general action is that to be expected from a therapy that is so effective in the treatment of

[194] I.A. Zakharova et al., "Endovascular Laser Therapy of Primary Glaucoma [Russian]," Vestnik Oftalmologii 110 (1994), No. 2, pp. 7-8

[195] Sukhanova (1993), pp. 124-32

vascular disorders, there might be other, more subtle effects through a boosting of general resistance to disease as well.

Infertility

While the treatment of pelvic inflammatory disease and sexually transmitted diseases in women discussed in Chapter 7 clearly can reduce infertility, it is also of interest that BI has shown itself effective in the treatment of male excretory infertility arising from long-term inflammation of the prostate, seminal vesicles, and testicles. In a study of 50 men aged 21-39 who suffered from excretory infertility, Sverdlovsk researchers formed a group of 25 who received standard therapy plus an unspecified amount of UBI and a control group of 25 who received standard therapy plus autotransfusion without irradiation.

Standard therapy consisted of gonovaccines in mounting doses, antibiotics, desensibilizing and disintoxicating therapy, physiotherapy, vitamin therapy, the use of biostimulators, and massage of the prostate gland.

The researchers noted that the administration of UBI led to feelings in some patients of blood flowing into the head, warming of various parts of the skin, mild dizziness, sleepiness, and lassitude during the procedure.

The UBI group were released from hospitalization 5-7 days sooner than the controls, and in some of the control group there was no improvement in sleep and appetite, which all the UBI group experienced. Tests following therapy revealed that the UBI group had less oligospermia and higher numbers of motile sperm. 10 pregnancies occurred subsequent to therapy in the UBI group, while 6 occurred in the control group.[196]

---

[196] A. P. Tarinskii et al., "Treatment of Male Excretory Infertility with UBI [Russian]," Akusherstvo i Ginekologiia (1990), No. 6, pp. 61-62

Miscellaneous

Various authors have reported that BI can treat impotence, repigment skin and hair, serve as a tonic in geriatric cases, treat chronic sinusitis, rhinitis, and laryngitis, and clear up mouth ulcers and gingivitis. It has been used repeatedly with excellent results in the treatment of shock. BI's immunosuppressive property makes it a strong candidate for use in Graft-Versus-Host-Disease following transplants. It is currently in use in the US against Lyme Disease and Chronic Fatigue Syndrome, but no results have been reported. Other indications in which BI has been found effective include tetanus, scarlet fever, hemophilia, Paget's disease of the bone, and delayed union of fractures.

In short, BI has been used in a very wide range of disorders with considerable effect, and in many with good to excellent effects.

Chapter 10: The Future of BI

BI therapy is healing with light, a most natural and appealing source. Even when it relies on artificial sources of that light, the central role of the secondary radiation from the irradiated cells in the mechanism of BI means that the pharmacologically effective source of light is a natural one--the capacity of the body's own cells to generate biophotons. These biophotons are channeled with unerring aim by the body's energy gradient to the exact locations where they can do the most therapeutic good.

This is a concept of consummate attractiveness. BI connects the energy of the macrocosm with the energy of the microcosm to create a highly effective yet unusually gentle pattern of action. It is a wonder of nature that this should be so, and it is not surprising that many of those who first come across reports of BI are inclined to skepticism. Yet the evidence in this book points overwhelmingly in the direction of the view that indeed Blood Irradiation therapy is very effective and safe when used in the appropriate manner.

Five main conclusions emerge from this account of the remarkable history and even more remarkable science of Blood Irradiation therapy:

1. BI therapy offers a curious example of a scientific advance that was largely ignored in the country of its origin even though it offered an excellent solution to serious, pressing problems and was promoted in a sober, professional manner. Then it came to be adopted in distant countries and developed to a high degree of precision amid widespread use. Yet such was the prejudice in its country of origin against the adopting countries that the very fact that they had

adopted BI became a mark against it. This resistance persisted even in the face of new pressing needs--the HIV/AIDS epidemic and the development of microbial resistance to antibiotics.

2. There are truly no serious critics of BI therapy, and there never have been any. What would a serious critic be like? He or she would take the trouble to investigate in depth the photobiology and pharmacology underlying BI; read widely in the medical literature in several languages; consult with the leading BI practitioners; and organize clinical trials that were designed to elucidate the strengths and weaknesses of BI head-on-head against the leading current treatments of various important diseases. None of those who criticize BI today has undertaken *even a single one* of these qualifying tasks. In other words, the irony is that this therapy that has so often been dismissed out of hand in fact does not have any serious critics to contend with. Perhaps the best source of one would be a BI practitioner who came to see that actually it was a useless or damaging therapy, but the fact that no such critic has emerged is telling.

3. Not only are there no serious critics; there also are no serious criticisms of BI therapy. Aside from the minor inconveniences of blood transfusion, a very modest level of toxicity, several dozen indications on which is it partially or totally ineffective, and the defects of the individual devices, meticulous investigation of every possible drawback of BI by many highly skilled researchers has identified not one significant problem. And repeated studies have shown that BI operates by a physiological mechanism that goes well beyond any placebo effect, though a placebo-like effect often accompanies BI.

4. BI is of unusual scientific interest. It offers insight into many aspects of pharmacology and photobiology. It can help elucidate the mechanisms of the body's energy metabolism. And its patterns of action can aid us in understanding related therapies as well.

5. BI therapy is highly effective, safe, and versatile. It has been exhaustively studied, and its mechanisms of action are well characterized. BI has been successfully used on millions of patients in a number of countries, and in various difficult indications for which there is no other effective treatment.

All of this augurs a bright future for this most efficacious yet mild of therapies--the therapy of the Healing Photons.

Appendix A: Side Effects

Although BI therapy has unusually low toxicity and in most cases causes no side effects at all, in some cases minor ones do occur. The most common side effect is a flushing of the skin, usually immediately after the reinfusion of the blood. The American pioneers found that this happened in up to 50 percent of their patients, but that may have been because they were treating a higher volume of blood under the Knott Technique than Germans and Russians subsequently have done.

The Americans ascribed the flushing to the oxygenation effect. An East German commentator considered that BI released histamine and prostaglandins, which in turn caused the flushing. It occurred in 5-15 percent of his patients, mostly in the head, and occasionally occurred several hours after the first treatment. This reaction was sometimes accompanied by a metallic taste, and in rare instances by the appearance of flecks caused by platelet aggregates in the eyes.[197]

Various practitioners have reported occasional feelings of fatigue, of warmth, and (rarely) of headache, sleep disturbance, hyperthyroid symptoms, and heightened sensitivity following BI. Exacerbations of bronchial asthma, allergies, and rheumatoid arthritis can occur following the first or second treatment. Catecholamines can be used in these cases.

BI can activate latent inflammation, leading to pain in abscessed teeth and the flare-up of latent appendicitis, cholecystitis, osteomyelitis, gastritis, ulcers, splinters, and scars. When BI dissolves platelet aggregates, embolization can occur in the legs. Frick advises to treat more frequently with BI, even twice daily, in these cases and says that at times surgical intervention may be required. Similarly, BI can occasionally stimulate menorrhea at the time of early menopause. In

[197] Frick (1989), p. 53. The following discussion draws on Frick, pp. 53-56

two cases herpes zoster activated following BI. The correct procedure in such cases is to treat with more BI to wipe out the activated virus. In general, when viral or immune cell activation follows BI, it is a sign that the disease is responsive and that higher doses will destroy the virus or excessively activated immune cells.

Diabetics can have hypoglycemic reactions following BI, so it is advisable for them to have carbohydrates just before or after treatment.

In patients with depression, the mood can worsen following therapy. In patients with labile personalities, the desire for more BI treatment can gain an addictive edge.

Of Frick's sample of 550 patients, 84 had side effects (15.3 percent). He himself notes that many of these would have occurred following placebo. In fact, it is clear that he was bending over backwards to note everything negative that occurred to these patients following UBI, regardless of whether it was connected to BI or not. For instance, the first registered side effect on his list is a suicide, with the comment that it followed the death of a son. In addition, Frick had deliberately chosen to employ UBI on patients who were middle-aged or older out of concern that there might be some carcinogenic result of therapy. Thus his patients were 53.7 years old on average. This means that both his positive results and his total side effects were skewed in a negative way because he was not treating younger, more resilient patients. His 15.3 percent total of side effects is far above those reported by other physicians and should be viewed as a worst-case estimate. Indeed, he himself was well aware of how low UBI's toxicity is in comparison with competing chemotherapies.

The American experience was summed up by Miley.[198] In 10,000 procedures he noted six adverse reactions attributed to the citrate he was using as an anticoagulant. They included headache, chill, temporary low fever, and moderate gastrocnemius spasm.

Miley found that UBI could safely be used after photosensitive drugs (e.g., sulfa drugs) had been discontinued, but that the use of photosensitive drugs following UBI was unpredictable and could be extremely dangerous and even fatal. He recommended waiting 4-7 days following UBI therapy before using photosensitive drugs, including digitalis.

In cases of widespread bacterial infection, including septicemia, following BI there can be a reaction arising from the sudden death of so many microorganisms. This reaction includes chill, a rise in

---

[198] (1997), pp. 21-23

temperature, and various toxic symptoms, all of which subside after a few hours.

The creation of a small amount of ozone at the time of the initial oxygenation of the blood following reinfusion of the irradiated blood cells is not known to do any damage to normal cells.

Incorrect technique and faulty veins can cause problems typical of all transfusion medicine.

Miley reported one extreme, fatal reaction following UBI that he considered possibly attributable to improper cleaning of the device, which allowed a piece of old dried fibrin to enter the blood. He noted that this was the only such instance reported in hundreds of thousands of treatments.

In regard to the possibility that the BI could cause an increase in mutagenicity that could lead to cancer, there is no evidence whatsoever that this has ever occurred. An East German study of the effects of 6 sessions of UBI found that in fact the number of chromosome aberrations dropped following therapy, perhaps because it stimulated the process of DNA repair.[199]

The very low level of irradiation in BI, the relatively rapid turnover of blood cell populations, and the ever-changing selection of cells for direct irradiation in repeated sessions of BI all argue against increased mutagenicity. Indeed, both the mechanism of BI's action in suppressing excessive metabolic activity (in this case, of incipient tumor cells) and anecdotal evidence regarding the very low rate of new cancers in former BI patients suggest that BI has a powerful prophylactic action against cancer similar to that of aspirin against colon cancer.

Conclusion

It is important to view this list of potential side effects of BI in proper perspective. In comparison with the toxicities of competing chemotherapies, the side effects in the above list tend to be less frequent and milder. There are reasons to believe that BI's pattern of toxicity resembles that of nucleoside analogues, as was discussed in Chapter 3, though BI's greater specificity means that toxic symptoms only emerge after a relatively higher dose. In normal usage, toxicity does not

[199] Frick (1989), pp. 55-6

represent a problem. Not only does this make BI a very safe therapy; it also allows experienced practitioners to use BI to the greatest effect without fear that it will cause unexpected damage.

Those knowledgeable in the field of toxicology will recognize that the very fact that a therapy's side effects can be enumerated, defined, and related to those of competing therapies actually undercuts the suspicion that it might have some hidden toxicity. One of the weaknesses in the way BI has at times been promoted has been the assertion that it has no toxicity at all. That is not believable in a therapy that can have such powerful therapeutic action as BI does. It is equally not believable, however, that a therapy that has such meticulously recorded side effects as those listed above also has some hidden damaging effect. Hundreds of BI practitioners desirous of ensuring their patients' well-being and protecting their own reputations have observed the results of BI therapy with eagle eyes. That in millions of cases over 70 years they have found no hidden damaging effect should tell any reasonable observer that it is extremely unlikely that such an effect exists.

Appendix B: The Knott Hemo-Irradiator

The Knott Hemo-Irradiator first patented by Edblom and Knott in 1928 (U.S. Patent No. 1,683,877--see below) was a largish apparatus that used a water-cooled mercury-quartz lamp. Its spectral distribution ranged from UV-C into the visible range, with a peak in UV-A. The intensities in the ultraviolet wavelengths were:

Wavelengths (nm)	*Milliwatts per square centimeter*
239.9	55,000
248.3	88,000
253.6	273,000
265.2	239,000
275.3	86,000
280.4	139,000
289.7	96,000
296.7	224,000
302.2	435,000
312.5-313.2	718,000
334.2	73,000
347.3	41,000
365.0-365.4	1,538,000

Sept. 11, 1928. 1,683,877

L. A. EDBLOM ET AL

MEANS FOR TREATING BLOOD STREAM INFECTIONS

Filed March 3, 1927

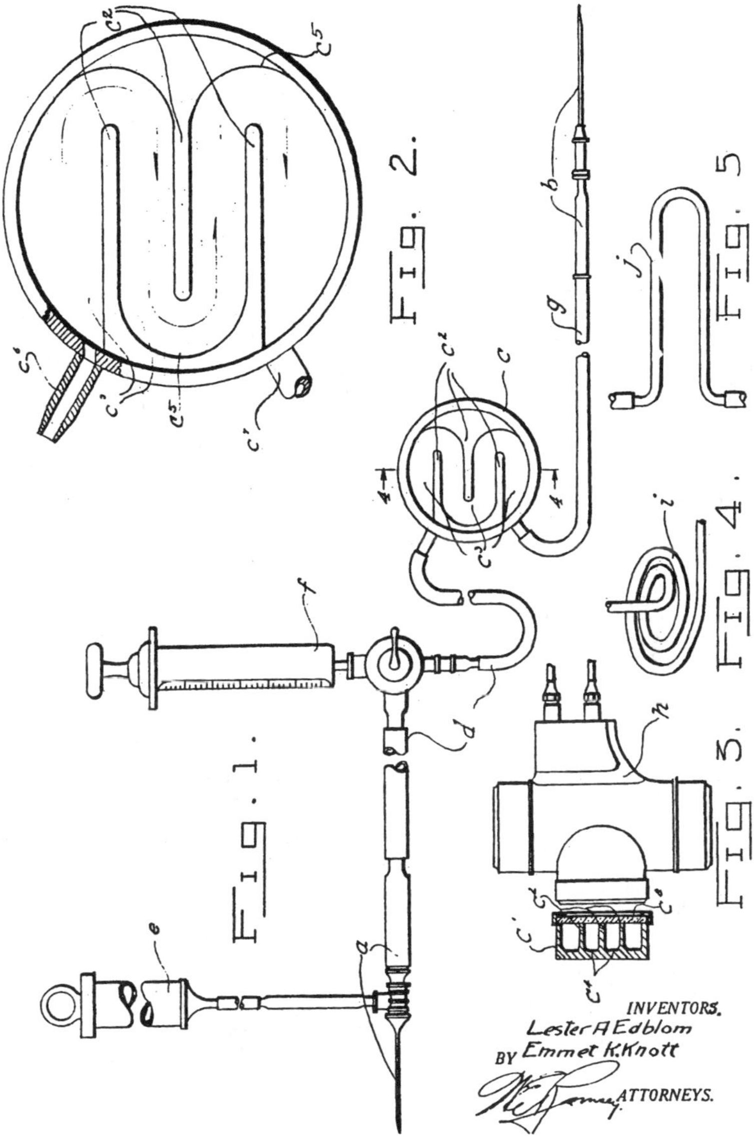

INVENTORS.
Lester A Edblom
BY Emmet K. Knott

ATTORNEYS.

Patented Sept. 11, 1928. **1,683,877**

UNITED STATES PATENT OFFICE.

LESTER A. EDBLOM, AND EMMET K. KNOTT, OF PORTLAND, OREGON.

MEANS FOR TREATING BLOOD-STREAM INFECTIONS.

Application filed March 3, 1927. Serial No. 172.358.

Our invention relates to the art of surgery and medicine and has for its object the provision of a novel method and means for treating blood stream infections and other pathologic blood conditions, and to this end is effective to destroy infectious bacteria, to stimulate normal blood cells and to inactivate toxic serum, by the use of ultra violet rays.

These phenomena of ultra violet radiation are generally known and are utilized in the treatment of external infections, and one of the principal objects of our invention therefore, is to use effectively these phenomena in the treatment of blood stream infections. We have discovered further that all toxins in the blood stream that we have exposed to ultra violet radiation, have been inactivated.

The treatment of blood in the circulatory system is rendered difficult because all of the blood vessels are encased in a layer of tissue constituting the walls of the vessel, and such layer of tissue is not permeable to the range of ultra violet radiation having bactericidal properties. When bacteria have thus entered the blood stream or become otherwise lodged under even the thinnest layer of tissue, they are protected against the bactericidal effect of radiation and can be destroyed only by the indirect manner of stimulating normal cells to combat such infectious organisms by their increased activity.

Clinical experiments have further determined that beneficial energy is stored up in the rayed blood temporarily, and if such blood can be returned to the blood stream immediately after it has been rayed, it will throw off secondary radiations which will stimulate and energize the patient.

We attain our object in a practical manner by tapping one of the patient's blood vessels, removing a small portion of blood in a closed air-tight conduit having one section permeable to the ultra violet rays possessing bactericidal power, and by forcing such blood thru such section while it is exposed to ultra violet radiation of the lengths mentioned, then returning the rayed blood immediately to another vessel in the body. This process is preferably carried on continuously over a period of time so that a predetermined portion of blood is rayed and rendered bacteria-free, and in this manner a large amount of blood can be withdrawn and rayed without lowering the patient's vitality to a very great degree. The portion rayed at one operation is determined by the patient's vitality, his power to throw off the organisms thus killed and remaining as foreign proteins in the blood stream, and by the condition to be corrected. The operation is repeated at intervals until a sufficient number of the infectious bacteria have been killed, sufficient normal cells have been stimulated, and/or a sufficient amount of toxic serum has been inactivated, to permit the normal bodily functions to take care of the pathological condition without further external aid.

The preferred apparatus for utilizing this method comprises a veni-puncture needle which is inserted into a vein, the butt end of the needle being attached to a piece of tubing of required length into which a container of citrate or oxylate solution is connected. The other end of such tubing is attached to a transfusion pump, which is connected at its outlet side to an exposure chamber covered by a quartz lens or other substance permeable to ultra violet rays. A generator of rays between 1800–4000 A. U. is arranged to direct such rays thru such quartz lens upon the blood stream. Such exposure chamber is constructed to produce turbulence to prevent setting out of the constituent parts of the blood, and the uniform and thorough raying thereof.

The outlet end of such chamber is connected by tubing to another veni-puncture needle which is inserted into another vessel of the same patient. If the method is used in blood transfusion to insure the purity of the blood, the second veni-puncture needle would, of course, be inserted into the patient to whom the blood was being given. The operation is performed in a room in which the temperature is maintained at such heat as to prevent a change in temperature of the blood being rayed, the apparatus is thoroughly sterilized, and great care is taken to prevent any air from entering the closed circuit.

The different bacteria to be destroyed contain various amounts of the photo-sensitive amino acids mentioned, and it is desirable to force the blood being treated thru the exposure chamber in the shortest possible time and thus each blood stream infection has its optimum exposure period. During this period the infectious bacteria are coagulated by the chemical action of the ultra violet rays and normal cells are stimulated by the longer length rays to better perform their functions. Over-stimulation of such cells causes their ultimate destruction but happily, observation has proven, there is a wide gap between the time of chemical destruction of foreign bacteria and the impairment or physical breakdown of normal cells due to over stimulation, somewhat in the relation of five seconds to two hundred and forty-five seconds.

The details of our improved method and the means in which it is embodied and efficiently performed, are hereinafter described in greater detail with reference to the accompanying drawings, in which:

Fig. 1 is a diagrammatic representation of the mechanism used, such devices comprising the latter being shown connected together;

Fig. 2 is a larger scaled plan view of the exposure chamber with the quartz lens removed;

Fig. 3 is a cross-section of such exposure chamber, and shown in relation to a quartz lamp; and

Figs. 4 and 5 are alternative forms of exposure chambers which might be used but are not considered as desirable as the one shown in the other figures.

The devices used are two veni-puncture needles *a* and *b*, the needle *a* serving as the inlet end of the closed conduit and the needle *b* as the outlet end. The butt end of the needle *a* is connected to the exposure chamber *c* by a piece of tubing *d*, preferably rubber, to provide the desired flexibility. Connected into such tubing is a container *e* for solution, citrate or oxylate, which solution is introduced into the blood stream to prevent coagulation and is a common expedient in blood transfusion technique. A second piece of tubing *g* connects the outlet end of the exposure chamber *c* to the needle *b*.

A force pump *f*, of the type used in blood transfusion also is connected in such tubing *d* so as to move such blood thru the exposure at a predetermined rate of flow.

The exposure chamber *c* comprises a cup-shaped body *c'* having a plurality of laterally extending, substantially parallel baffle walls c^2 defining a tortuous passageway of uniform cross-section from the inlet c^6 to the outlet c^7 of such exposure chamber. Such exposure chamber *c* preferably is set up on its edge with the baffle walls extending in a horizontal plane. The walls are spaced alternately from the opposite sides of the body, and thus when blood is forced thru such body, it is required to flow over the ends of such baffles and to be agitated thereby. Inasmuch as the walls are thus arranged in a vertical stack the blood flows in a continuous descending path and thus the cells are not permitted to settle out and to form thrombi. As is shown in Fig. 3, the connection between the walls and the bottom of the cup-shaped body is filleted, as at c^4, to eliminate a sharp corner in which particles might become lodged, and subsequently become loosened and carried along by the blood stream. The ends of the baffle walls c^2 are connected by rounded portions c^5 also to eliminate this possibility of thrombi formation, and also to eliminate a sudden corner around which the force of the flow might whip out the solid particles carried by the stream, and thus destroy the nice uniform mixture of the latter.

The entire chamber, or at least the portions thereof in contact with the blood stream, is made of a material with which the blood will not react chemically, such for example as silver or glass. The passageway also is of uniform depth which depth is less than the depth of effective radiation and preferably is approximately one centimeter.

The tortuous passageway c^3 causes the blood to be gently agitated so that the turbulence thus created turns all of the particles carried in suspension to the ultra violet rays so that the latter will be enabled to effect its beneficial influence on all of the blood cells and its lethal influence upon infectious bacteria carried by such stream.

A lens of quartz c^8, or of other material permeable to ultra violet radiation, is fitted air-tight upon the open top c^9 of the exposure chamber and thru the lens the ultra violet rays are directed by a generator, preferably of the water-cooled type capable of delivering rays between 1800–3000 A. U. and having a large proportion of such rays in the region of 2540 A. U.

There may be one or more of such exposure chambers *c* and generators *h* connected in series or in parallel, to produce the ideal condition to the blood leaving the circulatory system and being returned thereto bacteria-free, stimulated and with no active toxins present but physically unchanged except diluted slightly, and to effect this result in the shortest possible time and with the least amount of blood withdrawn from the circulatory system at one time.

It is to be understood also that altho the particular form of exposure chamber is preferable, it is not essential. A straight or flat spiral, or a reversed tube, as shown as *i* and *j*, respectively, in Figs. 4 and 5, can be used altho they are not believed to be as practical as the one shown in Figs. 1 to 3, because of their increased cost and frailty, and difficulty of manufacture and because of a less efficient mixture of the blood stream attained, while the latter is passing under the influence of the beneficial ray. It is believed essential in every exposure chamber that the bore be smooth but that means be provided for causing the cells to be agitated gently so that all cells and bacteria be exposed to the ultra violet radiation.

In our experimentations we have used an exposure chamber of the type shown in Figs. 1 to 3, inclusive, in which the passageway is approximately 1 square centimeter in cross-sectional area, and about 1 centimeter deep, and a flat quartz lens approximately one-quarter centimeter thick was sealed tightly over the chamber. A water cooled type Uviarc quartz tube, located as close as possible, was used as the generator of the ultra violet radiation, the generator being operated at 60 volts, and approximately 4 amperes, and the room temperature was maintained at about 90 degrees F.

Under such conditions, small quantities of human blood in which cultures of Staph. aureus were mixed, were rayed for the following exposures: 0—5—10—15—20—30—35—45—60—70—80—90—100—120—140—160—180—220—260—280 seconds. Cultures obtained, taken after each exposure showed growth only on the first three exposures, and no growth after 10 seconds exposure.

There was no change in the forms of th blood cells up to 70 seconds exposure, then slight poiklocytosis noted which condition disappeared after 80 seconds. The cells were normal and no change was noted until 280 seconds then disintegration of a mild degree was noted. Achromogenises also noted at 280 seconds.

We claim:

1. Apparatus of the character described for treating blood stream infections, comprising in combination with a generator of ultra violet radiation between 1800–4000 A. U., an airtight conduit having veni-puncture needles at the ends thereof, an exposure chamber connected in said conduit, intermediate such ends, a pump element connected to said conduit and adapted to cause flow therethrough at a predetermined rate, such exposure chamber having a portion permeable to ultra violet radiation, and such portion being arranged relatively to such generator as to permit the rays directed from the latter to pass therethrough.

2. Apparatus of the character described for treating blood stream infections, comprising in combination with a generator of ultra violet radiation between 1800–4000 A. U., an airtight conduit having veni-puncture needles at the ends thereof, an exposure chamber connected in said conduit, intermediate such ends, a pump element connected to said conduit and adapted to cause flow therethrough at a predetermined rate, such exposure chamber having a portion permeable to ultra violet radiation, and such portion being arranged relatively to such generator as to permit the rays directed from the latter to pass therethrough, and provided with means for producing turbulence in the blood passing by such permeable portion, for the purpose specified.

3. Apparatus of the character described for treating blood stream infections, comprising in combination with a generator of ultra violet radiation between 1800–4000 A. U., an airtight conduit having veni-puncture needles at the ends thereof, an exposure chamber connected in said conduit, intermediate such ends, a pump element connected to said conduit and adapted to cause flow therethrough at a predetermined rate, such exposure chamber having a portion permeable to ultra violet radiation, and such portion being arranged relatively to such generator as to permit the rays directed from the latter to pass therethrough, and the portions of the conduit in the exposure chamber being formed to produce turbulence in the blood passing by such permeable portion, for the purpose specified.

4. Apparatus of the character described for treating blood stream infections, comprising in combination with a generator of ultra violet radiation between 1800–4000 A. U., an airtight conduit having veni-puncture needles at the ends thereof, an exposure chamber connected in said conduit, intermediate such ends, means for supplying a diluting medium into such conduit arranged adjacent said exposure chamber, a pump element connected to said conduit and adapted to cause flow therethrough at a predetermined rate, such exposure chamber having a portion permeable to ultra violet radiation, and such portion being arranged relatively to such generator as to permit the rays directed from the latter to pass therethrough and provided with means for producing turbulence in the blood passing by such permeable portion, for the purpose specified.

5. Apparatus of the character described for treating blood stream infections, comprising in combination with a generator of ultra violet radiation between 1800–4000 A. U., an airtight conduit having veni-puncture needles at the ends thereof, an exposure chamber connected in said conduit, intermediate such ends, a container of diluting material connected into such conduit adjacent said exposure chamber, a pump element connected to said conduit and adapted to cause flow therethrough at a predetermined rate, such exposure chamber having a portion permeable to ultra violet radiation, and such portion being arranged relatively to such generator as to permit the rays directed from the latter to pass therethrough, and provided with side walls formed to produce turbulence in the blood passing by such permeable portion, for the purpose specified.

6. An exposure chamber of the character described, comprising a body defining a tortuous passageway of substantially uniform cross-section, one portion of said body separating the passageway from the exterior thereof being made of a material permeable to ultra violet radiation of 1800–4000 A. U.

7. An exposure chamber of the character described, comprising a body defining a tortuous smooth-bored passageway of substantially uniform cross-section, and descending progressively in a vertical plane, one portion

of said body separating the passageway from the exterior thereof being made of a material permeable to ultra violet radiation of 1800–4000 A. U.

8. An exposure chamber of the character described, comprising a body defining a tortuous passageway of substantially uniform cross-section, one portion of said body separating the passageway from the exterior thereof being made of quartz permeable to ultra violet radiation of 1800–4000 A. U.

9. An exposure chamber of the character described, comprising a body defining a tortuous smooth-bored passageway descending progressively in a vertical plane, one portion of the latter separating the passageway from the exterior thereof being made of a material permeable to ultra violet radiation of 1800–4000 A. U.

10. In apparatus for treating blood stream infections, in combination with devices adapted to be connected with a blood vessel of a patient to be treated including a tubular conduit having a section permeable to ultra violet radiation, means for pumping blood at a predetermined rate from said vessel thru such section, and means for directing ultra violet radiation of 1800–4000 A. U. upon the blood passing thru such section.

11. In apparatus for treating blood stream infections, airtight withdrawing and containing devices adapted to remove from the circulatory system a predetermined quantity of blood, means for exposing the same to radiation of 1800–4000 A. U. and returning such rayed blood to the circulatory system.

LESTER A. EDBLOM.
EMMET K. KNOTT

Appendix C: UBI Technique

The current Russian Isolda UBI device can be used according to the following technique.[200]

The accepted standard of UBI treatment is to irradiate a small portion of the blood for a limited amount of time and to repeat this treatment at intervals that are appropriate for the disease and its intensity, e.g., 5-6 sessions spaced one week apart to treat a moderate case of chronic hepatitis B. It is thought that in this way any danger that might arise from irradiating the total volume of the blood or irradiating the blood with a higher intensity UV source or for a longer time can be obviated. Clearly, in a fulminating condition, it might be necessary to use a higher dosage and to repeat the normal treatment at very frequent intervals to save the patient's life.

The standard procedure (Knott Technique) is to withdraw 1.5 ml of blood per pound of body weight (up to a total of 250 ml) by venipuncture. The blood flows into a transfusion flask containing anticoagulant (heparin, citrate, or saline). The blood is pumped through tubing at an automatically controlled rate. The blood flows through an irradiation chamber (cuvette) where it is exposed for up to ten seconds to a controlled amount of ultraviolet irradiation in the accepted therapeutic UV bandwidth. When the correct amount has been irradiated and stored in the flask, the direction is reversed and the blood is irradiated a second time on its passage back into the body through the same needle used for withdrawal.

Using gravity feed, the procedure takes about one hour including ten minutes for set-up and ten minutes for clean-up. Using the pump cuts the total time to about 30 minutes. The UV lamp should be turned on for 5-10 minutes to allow it to warm up.

---

[200] The photograph of the Isolda device on the following page is the Author's.

For a case that will requite repeated treatments, the patient's name can be etched on the crystal cuvette and the cuvette can be cleaned and sterilized. For single treatments, disposable cuvettes are also available. The accessories (tubing and needles) are disposed of after each use.

Figure 2. The Isolda Device

Technical Features of the Isolda UBI device

- Power of radiation source: 8 watts

- 80 percent of the UV irradiation is in the 254 nm region (UV-C). At 254 nm, the flux radiation area density within the cuvette plane varies from 2 to 3 milliwatts per square centimeter. At 308 nm, 313 nm, 365 nm, 408 nm, and 546 nm, the radiation flux area density is 0.1 to 0.4 milliwatts per square centimeter.

- The peristaltic pump operates at two speeds: 12 +/- 2 ml/min rate of flow; and 18 +/- 2 ml/min rate of flow.

- Volume of the irradiated blood: 1.5 ml per pound of body weight (approximately 5 percent of blood volume)

- Duration of procedure: 40 minutes

- Dimensions: 330 x 135 x 106 mm

- Weight: 4.4 kg

- Voltage: 220

Appendix D: Solar Blood Irradiation

Upon observing that both UBI and LBI had excellent effects and yet were not quite natural therapies, Spectrum Bioscience conceived of the notion of using the sun itself as the source of irradiation of blood cells. This Solar Blood Irradiation (SBI) would be a completely natural therapy that would, it is thought, be appealing to those who believed in natural healing. SBI would also be useful in remote and wilderness settings where no electricity was available as well as during the brownouts and blackouts common in the developing world, in natural catastrophes, or in wartime.

In addition, for certain patients, the sense of being treated with an entirely natural source and one possessed of the central iconic significance of the sun might convey a psychological effect that would boost the healing potential of SBI over those of UBI and LBI.

It was only later on, during the first steps in the development of the prototype of an SBI device, that Spectrum Bioscience came upon the results of the Vladivostok researchers reported in Chapters 3 and 5.[201] That UBI should be more effective than LBI would appear to result from the much broader spectrum of wavelengths it contains, which would more fully activate the blood cells. In turn, however, that suggested that SBI might be more effective than UBI because solar irradiation includes the entire spectrum.

A further step was to add a Full Spectrum Blood Irradiation (FSBI) capability to the SBI device by means of a full spectrum light that could be rotated out of the way during SBI. The full spectrum

---

[201] Sukhanova (1993)

FIGURE

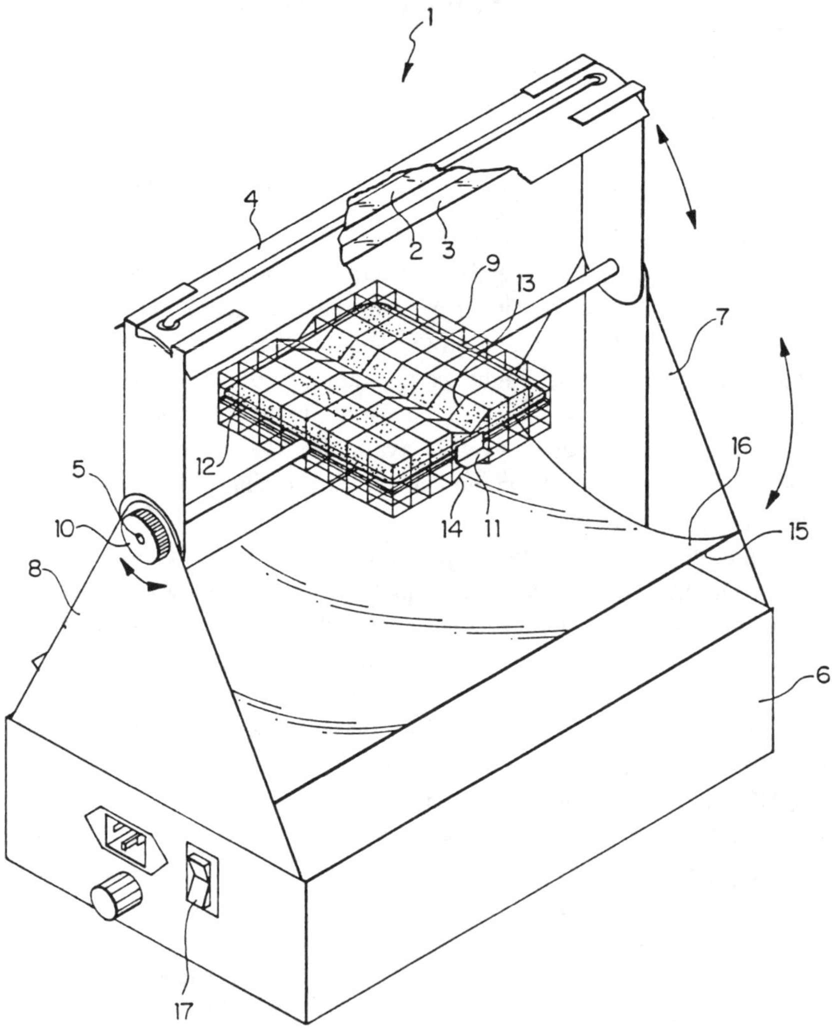

Figure 3. The Solar 1 Device

mode greatly extends the range of situations in which the device can be used. So the device came to have two modes--solar and full spectrum--and it came to be named Solar 1.

Spectrum Bioscience expects that the effect of the many wavelengths of solar light on the blood will be superior to those of UBI and LBI. This is, however, an hypothesis that will require testing to prove or disprove. The precise cause of SBI's superiority would require much study. It might involve such parameters as degree of polarization, coherence length, magnetic field, radiation pressure, and spectral distribution. But the overall sense would be that, just as UBI offers a more "complete" treatment of the irradiated cells than LBI, so too would SBI provide a more "complete" treatment than UBI (more wavelengths and a higher ability to draw a response from irradiated cells as a purely natural source to which the blood cells are finely attuned from primeval times).

Formally stated, this is a Solar Hypothesis of Blood Irradiation Therapy: the more closely a type of Blood Irradiation resembles solar irradiation, the higher its therapeutic effect. The varieties of BI could be arranged in ascending order thus: 1. photopheresis (lowest because it includes chemotherapy that diverges from the ideal of energy); 2. LBI (a single wavelength); 3. UBI (multiple UV wavelengths reminiscent of sunlight); 4. Full Spectrum BI; and 5. SBI (the ultimate, complete source).

It is worth noting that a Finnish crossover study found heliotherapy to be decisively superior to artificial ultraviolet irradiation in the treatment of psoriasis.[202] In addition, findings that pulsed irradiation is more effective than continuous and that molecular vibration occurs in irradiated cells even at wavelengths for which no spectral absorption is registered support the conclusion that the current First Law of Photochemistry (that irradiation has action only when absorbed) is not correct.[203] In turn, they buttress the Solar Hypothesis of Blood Irradiation Therapy.

Although Solar 1's solar mode will be dependent on latitude, weather, and time of day, it will presumably work under light cloud cover and thus will be functional in most places at most times of most days. At noon in the Sahara, 15 seconds of exposure of the blood might

---

[202] E. Snellman, "Comparison of the Antipsoriatic Efficacy of Heliotherapy and Ultraviolet B: A Cross-over Study," Photodermatology, Photoimmunology and Photomedicine 90 (1992), No. 2, pp. 83-85

[203] Ministry of Health of Ukraine (1996), pp. 125, 135

suffice. In the late afternoon of a cloudy August day in Winnipeg, it might take 15 minutes to reach the proper cumulative irradiation.

Spectrum Bioscience expects that certain practitioners and patients will strongly prefer an entirely natural therapy such as SBI. Solar 1 has the potential to become popular in settings such as sanatoria and natural medicine clinics as well as in remote locations. Its primary initial use, however, will be as an antiviral device. Potential applications would be early-stage HIV, chronic hepatitis, and overwhelming viral infections. Spectrum Bioscience is currently working with organizations that could apply Solar 1 to the treatment of infectious diseases in developing countries.

It is possible that certain patients who ordinarily do not respond to placebos will respond exceptionally well to SBI because they will perceive it as a quintessential natural therapy that is highly effective and very safe. In turn, other patients may be encouraged to seek out SBI treatment in the hope that it will have this effect on them also. This means that SBI could convey therapeutic results that are significantly superior to those of other forms of BI as well as of competing chemotherapies and natural medicines.

Select Bibliography

Note: The leading scientific study of BI is I.E. Ganilina and K.A. Samoilova, eds., *Mekhanizmy vliianiia obluchennoi ul'trafioletovymi luchami krovi na organizm cheloveka i zhivotnykh* (Mechanisms of the Influence of Blood Irradiated with Ultraviolet Rays on the Organisms of Humans and Animals.). Leningrad: Nauka, 1986. Other books on UBI have been published in Sverdlovsk and Vladivostok. Books on laser medicine that contain information on LBI have been published in Izhevsk, Khabarovsk, Kiev, L'vov, Minsk, Moscow, Novosibirsk, Obninsk, Saratov, Tashkent, and Vladivostok.

*****

Barger, G. and E.K. Knott, "Blood: Ultraviolet Irradiation (Knott technic)," in: Otto Glasser, ed. *Medical Physics*, 3 vols. (Chicago: Year Book Publishers, 1944-1960, Vol. II (1950), pp. 132-6

Barrett, Henry A., "The Irradiation of Autotransfused Blood by Ultraviolet Spectral Energy. Results of Therapy in 110 Cases," Medical Clinics of North America 24 (1940), pp. 723-32

Bednarskii, A.S. et al., "The Use of Intravascular LBI in the Combination Therapy of Preeclampsia [Russian]," Akusherstvo i Ginekologiia 6 (1995), pp. 18-22

Berdichevskii, M. Ia. and E.M. Dashkovskaia, "Efficacy of the Complex Treatment of Cerebrovascular Disorders with Ultraviolet Irradiation of

Autologous Blood [Russian]," Zhurnal Nevropatologicheskoi Psikhiatrii Im. S.S. Korsakova 91 (1991), No. 1, pp. 75-78

Beroza, Morton. The Effect of Short Time Ultraviolet Irradiation on Blood and Biochemical Compounds. Masters Thesis, Georgetown University, May 1946

Bischof, Marco. Biophotonen. Das Licht in unseren Zellen. Frankfurt: Zweitausendeins, 1995

Brill', G.E., ed. The Experimental and Clinical Use of Low-Intensity Lasers and Irradiation in the Millimeter Range [Russian]. Saratov, 1994

Coohill, Thomas P., "Action Spectra Again?" Photochemistry and Photobiology 54 (1991), No. 5, pp. 859-70

Cooke, M.A., "Secondary Emanatory Phenomena in Blood of Irradiated Subjects," British Journal of Physical Medicine 9 (1946), p. 172

Dillon, Kenneth J. Apprentice to Paracelsus: My Search for the Secrets of Healing. McLean, Virginia: McLean Research Associates, 1994

Douglass, William C. Into the Light. 2nd edition. Atlanta: Second Opinion, 1996

Duggar, B.M., ed. Biological Effects of Radiation. 2 vols. New York: McGraw, 1936

Eckart, W. et al. "Bericht ueber 2 Tagungen zu Fragen der biologischen Wirkung der UV-Strahlung (Brno 1972, Gorki 1973)," Zeitschrift fuer Physiotherapie 27 (1975), pp. 415-41

Edelson, R. "Light-activated Drugs," Scientific American 259 (August, 1988), No. 2, pp. 68-75

Edelson, R. et al., "Treatment of Cutaneous T-Cell Lymphoma by Extracorporeal Photochemotherapy," New England Journal of Medicine 316 (1987), pp. 297-303

Farmer, D.F. et al., "The Use of Hemo-Irradiation (Knott Technique) in Eye Infections," Industrial Medicine and Surgery 21 (1952), No. 4, pp. 183-5

Frick, G. Fibel der Ultraviolettbestrahlung des Blutes. Greifswald: Ernst-Moritz-Arndt-Universitaet Greifswald, 1989

Frolov, V.M. et al., "Compromised Immunity and Microcirculation and Their Correction in the Liquidators of the Aftermath of the Chernobyl' Reactor Accident, Presenting with Neurosomatic Pathologies [Russian]," Likarska Sprava (1995), No. 5-6, pp. 22-25

Hancock, V.K., "Treatment of Blood Stream Infections with Hemo-Irradiation. Case Reports," American Journal of Surgery 58 (1942), No. 3, pp. 336-44

Isakov, V.L. The Use of Lasers in Biology and Medicine [Russian and Ukrainian]. Kiev: Ukrainian Center of Laser Medicine, 1996

Jagger, J. Introduction to Research in Ultraviolet Photobiology. Englewood Cliffs NJ: Prentice-Hall, 1967

Knott, E.K., "Development of Ultraviolet Blood Irradiation," American Journal of Surgery 76 (1948), No. 2, pp. 165-71

Komarova, L.A. and G.I. Yegorova. The Medical Application of Low-Intensity Laser Irradiation [Russian]. St. Petersburg: St. Petersburg Academy for Postdiploma Medical Education, 1994

Kul'chavenia, E.V., "Use of Low-Intensity Lasers in Tuberculosis [Russian]," Problemy Tuberkuloza (1995), No. 4, pp. 19-21

Kutko, I.I., V.V. Pavlenko, and E.G. Voronkov, "The Use of Intravascular Laser Blood Irradiation in the Treatment of Forms of Schizophrenia Resistant to Therapy [Russian]," Zhurnal Nevropatologii i Psikhiatrii Imeni S.S. Korsakova 92 (1992), No. 4, pp. 53-56

Marchenko, A.V. et al., "The Possibility of Using UV-Irradiated Blood in the Prophylaxis of Serum Hepatitis during Hemotransfusion in

Surgery [Russian]," Vestnik Khirurgii Imeni I. I. Grekova (1990), No. 2, pp. 87-89

Marochkov, A.B. et al., "Complications in the Ultraviolet Irradiation of the Blood [Russian]," Anesteziologiia i Reanimatologiia (1990), No. 4, pp. 55-56

Matsuyev, A.I. et al., "The Use of Ultraviolet Blood Irradiation in Obstetrical-Gynecological Practice [Russian]," Akusherstvo i Ginekologiia (1990), No. 8, pp. 6-10

Miley, G.P., "Ultraviolet Irradiation of Auto-Transfused Human Blood, Studies in Oxygen Absorption Values," American Journal of Medical Sciences 197 (1939), No. 6, p. 873

Miley, G.P., "The Knott Technic of Ultraviolet Blood Irradiation in Acute Pyogenic Infections," N.Y. State Journal of Medicine 42 (1942), No. 1, pp. 38-46

Miley, G.P., "Efficacy of Ultraviolet Blood Irradiation Therapy in the Control of Staphylococcemias," American Journal of Surgery 64 (1944), No. 3, pp. 313-22

Miley, G.P., "Ultraviolet Blood Irradiation Therapy in Acute Poliomyelitis," Archives of Physical Therapy 25 (1944), pp. 651-6

Miley, G.P., "Recovery from Botulism Coma Following Ultraviolet Blood Irradiation (Knott Technic)," Review of Gastroenterology 13 (1946), No. 1, pp. 17-19

Miley, G.P. and J.A. Christensen, "Ultraviolet Blood Irradiation Therapy in Acute Virus and Virus-Like Infections," Review of Gastroenterology 15 (1948), No. 4, pp. 271-77

Miley, G.P. and J.A. Christensen, "Further Studies in Acute Infections. Ultraviolet Blood Irradiation Therapy," American Journal of Surgery 73 (1947), No. 4, pp. 486-93

Miley, G.P. and E.W. Rebbeck, "The Knott Technic of Ultraviolet Blood Irradiation as a Control of Infection in Peritonitis," Review of Gastroenterology 10 (1943), No. 1, pp. 1-28

Miley, G.P. et al., "Intractable Bronchial Asthma, Preliminary Report of Results Observed in 80 Cases," Archives of Physical Therapy 24 (1943), pp. 533-42

Miley, G.P. et al., "Ultraviolet Blood Irradiation Therapy of Apparently Intractable Bronchial Asthma," Archives of Physical Medicine 27 (1946), pp. 24-29

Miley, G.P., R.C. Olney, and H.T. Lewis. Ultraviolet Blood Irradiation: A History and Guide to Clinical Application (1933-1997). Silver Spring, Maryland: Foundation for Blood Irradiation, 1997

Moor, F.B. et al., "An Evaluation of Hemoirradiative Therapy," Archives of Physical Medicine 29 (1948), pp. 358-65

Olney, R.C., "Ultraviolet Blood Irradiation, Knott Method, in Biliary Disease," American Journal of Surgery 72 (1946), pp. 235-7

Olney, R.C., "Ultraviolet Blood Irradiation Treatment of Pelvic Cellulitis, Knott Method," American Journal of Surgery 84 (1947), No. 4, pp. 440-3

Olney, R.C. et al., "The Treatment of Viral Hepatitis with the Knott Technic of Blood Irradiation," American Journal of Surgery 90 (1955), No. 3, pp. 402-9

Ovsiannikov, V.A., "Analysis of the Low-Energy Laser Treatment of Some Cancers and Infectious Diseases," Journal of Clinical Laser Medicine & Surgery 15 (1997), No. 1, pp. 39-44

Rebbeck, E.W., "Ultraviolet Blood Irradiation in the Treatment of Postabortional Sepsis," American Journal of Surgery 82 (1942), No. 6, pp. 736-40

Rebbeck, E.W., "Preoperative Hemo-Irradiations," American Journal of Surgery 61 (1943), No. 2, pp. 259-65

Rebbeck. E.W., “Ultraviolet Blood Irradiation in the Treatment of Escherichia Coli Septicemia,” Archives of Physical Therapy 24 (1943), pp. 158-67 and 176

Rebbeck, E.W., “Use of Ultraviolet Blood Irradiation (Knott Technic) in Biliary Tract Surgery,” American Journal of Surgery 80 (1950), No. 1, pp. 106-12

Rebbeck, E.W. and H.T. Lewis, “The Use of Ultraviolet Blood Irradiation in Typhoid Fever,” Review of Gastroenterology 16 (1949), pp. 640-9

Schleicher, C., “Application of Ultraviolet Blood Irradiation for Treatment of HIV and Other Bloodborne Viruses,” Townsend Letter for Doctors and Patients 147 (1995), pp. 66-72

Schwartz, Steven O. et al., “Ultraviolet Irradiation of Blood in Man,” JAMA 149 (1952), pp. 1180-3

Snellman, E., “Comparison of the Antipsoriatic Efficacy of Heliotherapy and Ultraviolet B: a Cross-over Study,” Photodermatology, Photoimmunology, Photomedicine 9 (1992), pp. 83-85

Stulin, J.D. et al., “Cerebral Hemodynamics in Schizophrenics with Depressive Syndrome during Intravenous Laser Treatment [Russian],” Zhurnal Nevropatologii i Psikhiatrii Imeni S. S. Korsakova 94 (1994), No. 1, pp. 54-56

Sukhanova, G.I. ed. Laser Therapy in the Far East [Russian]. Vladivostok: Dal’nauka, 1993

Taubert, K., “Ultraviolettbestrahlung des Blutes bei der Migraene,” Zeitschrift Aerztlicher Fortbildung (Jena) 85 (1991), No. 1-2, pp. 43-48

Voeikov, V.L. et al., “Alterations in Luminol-enhanced Chemiluminescence from Nondiluted Whole Blood in the Course of Low-level Laser Therapy of Angina Pectoris Patients,” SPIE Proceedings (1997), Vol. 2895

Wiesner, Anke, "Zur Geschichte der UV-Therapie," Folia Haematologica, Leipzig 101 (1974), No. 5, pp. 848-56

Zmudzka, B. and J. Beer, "Activation of Human Immunodeficiency Virus by Ultraviolet Radiation," Photochemistry and Photobiology 52 (1990), No. 6, pp. 1153-62

Zvereva, K.V. et al., "The Choice of Method of Intravascular Laser Therapy in Rheumatoid Arthritis [Russian]," Terapevticheskii Arkhiv 66 (1994), No. 1, pp. 29-32

Index

A

B

C

D

E

F

G

H

N

O

P

Q

R

S

T